FLORA OF TROPICAL EAST AFRICA

GUTTIFERAE*

P. BAMPS, N. ROBSON & B. VERDCOURT**

Trees, shrubs, subshrubs or herbs, rarely climbers, mostly containing a yellow, orange or clear resinous latex, usually glabrous but occasionally with an indumentum of simple or stellate hairs. Leaves usually opposite, sometimes verticillate or alternate, simple, entire or very obscurely crenate, exstipulate, usually with glands and resin channels which are often translucent. Inflorescences terminal or axillary, 2-fid cymes, false racemes, panicles, fascicles or umbels or sometimes the flowers solitary. Flowers regular, often fleshy, hermaphrodite, dioecious or polygamous. Sepals (2–)4–5(–6 or more), imbricate, often decussate. Petals free, 4–5(–6 or more), imbricate or contorted, sometimes decussate. Stamens usually numerous, free or very often in fascicles or fused into groups, anthers usually with 2 thecae dehiscing by longitudinal slits; the outer fascicles are often sterile "fasciclodes", frequently forming what is often called the disc, or are absent. Ovary superior, 1–several-locular, with parietal, axile, apical or basal placentation; ovules anatropous, 1–∞ per locule; styles free or fused, sometimes very short or absent; stigmas or stigma-lobes mostly equal in number to the locules. Fruit a berry, drupe or septicidal (rarely loculicidal) capsule. Seeds arillate or not, without albumen, sometimes winged; embryo with cotyledons often very reduced.

Except for *Hypericum* L., which is very widely dispersed, principally a tropical family with 40 genera and about 1000 species.

Pentadesma butyracea Sabine has been cultivated, e.g. Tanzania, Lushoto District, Siri Plantation SC7, 7 Mar. 1932, *Greenway* 2939! & Uganda, Entebbe Botanic Garden *fide* Dale, Introd. Trees Uganda: 54 (1953). It is a W. African and Zaire species distinguished by being a large tree to 35 m. with flowers in racemiform cymes, petals glabrous on the inside, 5 stamen-fascicles, style simple and divided into 5 lobes, ovary 5-locular with 7–14 ovules in each locule and baccate fruits. The wood is useful and the seeds yield a vegetable butter. It is figured in F.W.T.A., ed. 2, 1, fig. 112 (1954), Aubrév., Fl. For. Côte d'Ivoire, ed. 2, 2: 326, t. 249 (1959) & Keay, Onochie & Stanfield, Nigerian Trees 1, fig. 47 (1960).

Styles 1–5, well developed; flowers hermaphrodite:
  Style single, with a peltate stigma; fruit ± perfectly
    globose, with 1 large seed; leaves with exceed-
    ingly numerous very close parallel lateral
    nerves; mainly a strand plant .   .   . 1. **Calophyllum**
Styles 2–5, free or ± joined; fruit several–many-
  seeded:
    Stamens joined to form a tube; styles joined,
      spreading at the apex; fruit a berry .   . 2. **Symphonia**

* The family Hypericaceae is now usually included in the Guttiferae but has already appeared as a separate part of this Flora. The family description given above covers the Hypericaceae and the key includes all the genera of the combined families which occur in the Flora area. Guttiferae and Clusiaceae are alternative names and we have preferred to use the one better known in East Africa.
** This account has been written by B. Verdcourt but is to a large extent based on those pre-published by P. Bamps and N. Robson for Flore du Congo and Flora Zambesiaca respectively, both of whom have read and corrected the typescript.

Stamens in bundles or free; styles free or sometimes partly to rarely wholly joined:
   Tall tree to 30 m.; petals glabrous on the inner
   face; fruit a berry (cultivated)   .   .   **Pentadesma**
                                                            (see note above)

   Herbs, subshrubs, shrubs or small trees:
      Petals yellow, often tinged red, glabrous on
      their inner faces; cymes biparous or
      corymbiform or flowers solitary; fruit
      a capsule or rarely a berry   .   .   6. **Hypericum**
      Petals white, velvety on their inner faces;
      cymes corymbiform or umbelliform or
      flowers in panicles; fruit a berry or
      drupe:
         Fruit a berry:
            Ovules 5–8 per locule; berry ± 25–40-
            seeded   .   .   .   .   7. **Vismia**
            Ovules 1 per locule; berry 5-seeded   .   8. **Psorospermum**
        Fruit drupaceous, with 5 pyrenes; ovules
            2–4 per locule   .   .   .   9. **Harungana**
Style obsolete, the stigma sessile or subsessile, peltate
   or lobed and spreading (except in *Garcinia*
   xanthochymus a cultivated species where narrowed ovary-apex is style-like); flowers dioecious
   or polygamous, less often hermaphrodite:
   Ovary incompletely 5-locular, with 2–28 ovules per
      locule; placentation parietal; stamens grouped
      in 5 fleshy bundles, the anthers subsessile and
      disposed in several rows superposed on the in-
      ner face or on both faces of the bundle; petals
      (1–)1·5–3·5 cm. long; fruits large, 10–34 × 6·5–
      17 cm.   .   .   .   .   .   3. **Allanblackia**
   Ovary completely 2–5-locular; ovules 1–2 per
      locule; placentation basal or apical:
      Fruit a drupe; ovary 2-locular, with 2 ovules per
         locule, or 4-locular, with 1 ovule per locule;
         placentation basal; stamens fused into a ring
         at the base; filaments well developed; calyx
         entire in bud, dividing into 2(–3) segments
         during anthesis; petals (1–)1·5–2 cm. long .   4. **Mammea**
      Fruit a berry; ovary 2–4-locular, with 1 ovule per
         locule; placentation apical; stamens free or
         inserted on a cylindrical to cupular disc or
         grouped in 2–5 bundles with the anthers
         borne at the summit, either on distinct free
         or connate filaments or ± sessile; calyx with
         (3–)4–5 distinct lobes in bud; petals 0·5–
         1(–1·2) cm. long in native species   .   .   5. **Garcinia**

1. CALOPHYLLUM

L., Sp. Pl.: 513 (1753) & Gen. Pl., ed. 5: 229 (1754); Henderson & Wyatt-
Smith in Gard. Bull., Singapore 15: 285–375 (1956); Stevens in Austral.
Journ. Bot. 22: 349–411 (1974)

   Trees or rarely shrubs, secreting a milky yellow or clear latex; bark of
young trees with characteristic diamond- or boat-shaped fissures. Leaves
opposite, petiolate or rarely ± sessile, entire, usually coriaceous, with very

numerous and closely placed fine parallel lateral nerves alternating with ± translucent glandular canals. Flowers terminal or axillary, in few–many-flowered racemes or panicle-like cymes or rarely reduced to 1–3, ♀. Sepals 4, free, the inner pair sometimes ± petaloid. Petals 2 or 4–8 or sometimes absent, white, imbricate, not always distinguishable from the inner sepals. Stamens numerous, usually arranged in 4 bundles opposite the petals or sometimes ± free; filaments slender and ± flexuous; anthers ovate to linear-oblong. Ovary 1-locular, with solitary erect ovule; style simple, slender, often flexuous; stigma peltate. Fruit a drupe, with brittle or thick pericarp and 1 large seed.

A large genus of about 140 species, predominantly in tropical Asia and Australasia, but also in Madagascar, Mascarene Is., eastern Africa and tropical America.

C. parviflorum Bak. is recorded from Zanzibar (U.O.P.Z.: 165 (1949)), but the description given does not agree at all with this species, which is characterised by its small leaves. Without material from Kizimbani Experimental Station it is not clear what is meant.

C. inophyllum *L.*, Sp. Pl. 1 : 513 (1753); P.O.A. B, fig. 15, 18 (1895); P.O.A C : 275 (1895); V.E. 3(2): 506 (1921); U.O.P.Z.: 163, fig. (1949); T.T.C.L.: 241 (1949); Perrier, Fl. Madag., Guttif.: 6 (1951); Henderson & Wyatt-Smith in Gard. Bull., Singapore 15 : 314, fig. 4/A, B (1956); Robson in F.Z. 1 : 394, t. 76 (1961); K.T.S.: 231 (1961); Stevens in Austral. Journ. Bot. 22 : 374 (1974). Type: Sri Lanka, near Colombo, *Hermann* (BM-HERM, 2 : 82, lecto. !)

Tree 7·5–30 m. tall, usually with a short trunk and long spreading branches; bark pale grey and fawn with shallow elliptic longitudinal fissures; branches smooth, 4-angled when young; branchlets ribbed and bearing obvious leaf scars when dry; leaf-blades bright green, elliptic-oblong to obovate, 8–18·5(–20) cm. long, (4·5–)5–12 cm. wide, rounded to slightly emarginate at the apex, broadly cuneate at the base, the margin ± undulate, coriaceous, the lateral nerves prominent on both surfaces; petiole 1–2·2 cm. long, broadened and flattened towards the apex. Inflorescences falsely racemose, lax, 3–12-flowered, 7–15 cm. long, in the upper axils; buds globose; pedicels 1·5–4 cm. long. Sepals 4, reflexed, deciduous, the outer pair round, 7–8 mm. long, the inner obovate, 1 cm. long, rounded at the apex, ± petaloid. Petals 4, obovate, 0·9–1·2 cm. long, narrower than the inner sepals, reflexed, deciduous. Stamens in 4 bundles, yellow or orange, equalling or rather shorter than the petals; anthers narrowly oblong, 1·5 mm. long. Ovary pink, globose; style ± 4 times as long as the ovary, slightly exceeding the stamens, flexuous. Fruit green, pale brown when dry, globose, 2·5–4 cm. in diameter, smooth, but coarsely wrinkled in dry state. Seeds brown, ovoid or subglobose, 1·7–2·2 cm. long and wide, ± mamillate, very oily. Fig. 1.

KENYA. Kilifi District: Shanzu Beach, 7 May 1967, *Greenway* 13093 !; Lamu District: Kipini, *Dale* !
TANZANIA. Pangani District: Mwera, Chengene, 24 May 1956, *Tanner* 2855 !; Rufiji District: Mafia I., Kisimani Mafia–Dundani, 27 Sept. 1937, *Greenway* 5327 !; Pemba I., near Finya, 12 Dec. 1930, *Greenway* 2712 ! & without exact locality, *Toms* 137 !
DISTR. **K**7; **T**3, 6; **Z**; **P**; Mozambique, Madagascar, Mascarene Is., tropical Asia and Malaysia extending to Melanesia and Polynesia, also cultivated in many parts of the world
HAB. Rocky and sandy sea-shores; 0–20 m., but also planted inland up to about 1200 m.

NOTE. The " Alexandrian laurel ", although undoubtedly often planted is certainly to be considered indigenous in some places along the coast.* The fruit is often washed up on beaches. It is probably often a relict of former cultivation for boat-building timber and is also frequently planted as an ornamental tree, e.g. Tanzania, Dar es

* A note by P. J. Greenway, whose experience of coastal E. Africa is unsurpassed, states on the Kew covers " the Zanzibar and Pemba specimens are wild and do not necessarily grow in or near native villages or huts ".

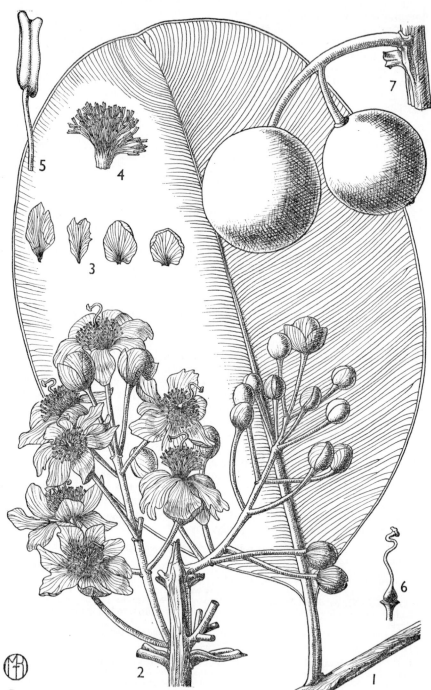

Fig. 1. *CALOPHYLLUM INOPHYLLUM*—**1**, leaf, × 1; **2**, inflorescence, × 1; **3**, petals, ×1; **4**, stamen-bundle, × 2; **5**, stamen, × 8; **6**, pistil, × 2; **7**, fruits, × 1. Drawn by Mrs. Olive Milne-Redhead.

Salaam Botanic Garden, 4 Oct. 1951, *Wigg* 965, Dar es Salaam, State House Grounds, 29 Aug. 1972, *Ruffo* 497, Pangani township, Sept. 1955, *Semsei* 2241, usually on the coast but also inland up to 300 m. or more; Dale, Introd. Trees Uganda: 16 (1953), records it from Entebbe Botanic Gardens and Kampala.

Fosberg has recently referred the populations of this species occurring in the western part of the Indo-Pacific (west of Sri Lanka) to a var. *takamaka* (K.B. 29: 255 (1974); type: Aldabra Atoll, South I., Takamaka Grove, *Fosberg* 49272 (US, holo., K, iso.!)) distinguished by having smaller fruits 1·7–2·5 cm. in diameter. I agree that there are differences, perhaps not as clear cut as Fosberg believes, but fruiting material from East Africa is too sparse for a proper assessment. I (B.V.) certainly remember drift fruits with dimensions near those of var. *inophyllum* and am content to follow P. F. Stevens who is monographing the genus and has annotated the Kew isotype " this is not worth recognising as a distinct variety ".

2. SYMPHONIA

L.f., Suppl. Pl.: 49 (1781)

Medium or large trees, rarely shrubs, with yellowish sap. Leaves opposite or subopposite, with numerous parallel lateral nerves running into a sub-marginal nerve; secretory canals not very visible. Flowers ♀, in sessile cymes, umbels or corymbs at the apices of short lateral branches or rarely solitary. Sepals 5, imbricate, persistent. Petals 5, contorted, deciduous. Disc cupuliform, ± 5-angled, persistent. Stamens joined into a tube sur-rounding the ovary, divided at the summit into groups of 2–6 anthers, de-ciduous. Ovary 5-locular, with 1–12 ovules in each locule; styles 5, joined at the base, then spreading-divergent at the apex, persistent. Berries with 1(–3) seeds. Seeds arillate with a thick entire embryo.

A genus of supposedly 17 species, 16 endemic to Madagascar and 1 common to America and tropical Africa. It is likely, however, that a few of the Madagascan " species " will prove to be conspecific with the following.

S. globulifera *L.f.*, Suppl. Pl.: 302 (1781); Oliv. in F.T.A. 1: 163 (1868); Engl. in Fl. Bras. 12: 469, t. 108 (1888); Vesque in DC., Monogr. Phan. 8: 227 (1893); V.E. 1(2): 635, fig. 545 (1910); V.E. 3(2): 519, fig. 234 (1921); Staner in Rev. Zool. Bot. Afr. 23: 221 (1933) & in B.J.B.B. 13: 143 (1934); Eyma in Fl. Suriname 3: 117 (1934); Lebrun, Ess. For. Rég. Mont. Congo Orient.: 161, t. 13 (1935); I.T.U., ed. 2: 155 (1952); F.W.T.A., ed. 2, 1: 292 (1954); Aubrév., Fl. For. Côte d'Ivoire, ed. 2, 2: 328, t. 250 (1959); Keay, Onochie & Stanfield, Nigerian Trees: 175, fig. 45 (1960); Robson in F.Z. 1: 394, t. 77 (1961); F.F.N.R.: 257 (1962); Bamps in F.C.B., Guttif.: 36, t. 4 (1970). Type: Surinam, *Dalberg* (not found*)

Glabrous tree (7·5–)15–25(–40) m. tall, with a straight cylindrical bole clear for 4·5–21 m., often with stilt roots surrounded by pneumatophores (in the New World); bark buff or greenish yellow or grey-brown, smooth or with vertical fissures and lenticels arranged in vertical rows; the wood when cut exudes a little yellow or orange sticky resin; branches horizontal, forming a rounded crown. Leaf-blades oblong, elliptic, oblanceolate, lanceolate, oblong-lanceolate or obovate-lanceolate, 5–12·5 cm. long, 1–5 cm. wide, obtusely acuminate at the apex, cuneate at the base, mostly very dark glossy green, ± coriaceous; petiole 0·5–2 cm. long, finely transversely wrinkled. In-florescences corymbose or subumbellate many-flowered cymes terminating short lateral branches; pedicels 0·4–2·6 cm. long, lengthening in fruit. Sepals round, reniform or ovate, 2–5 mm. long, 2·5–7 mm. wide. Petals waxy, crim-son or scarlet, round, 0·7–1·7 cm. long, 1·3–2 cm. wide, slightly spreading.

* It is not unlikely that specimen 853.1 in the Linnean Herbarium, bearing the single word *Symphonia*, could be this missing type; some of the information given in the description but not available from the specimen could have been given as a note by Dalberg.

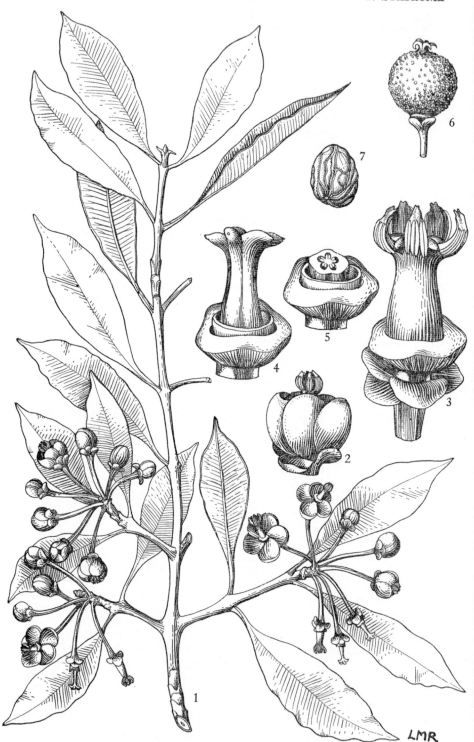

FIG. 2. *SYMPHONIA GLOBULIFERA*—1, flowering branch, × ⅔; 2, flower, × 2; 3, flower, petals removed, × 4; 4, flower, perianth and most of staminal tube removed, × 4; 5, same except ovary cut transversely, × 4; 6, fruit, ×1⅓; 7, seed, × 1⅓. 1–5, from *Milne-Redhead* 2948; 6, 7, from *Holmes* 1249. Reproduced with permission of the Editors of " Flora Zambesiaca ".

Disc 1·5–4 mm. tall, with entire or undulate border, ± pentagonal. Staminal tube 3·5–10 mm. tall, ± persistent after the petals fall; anthers linear, 3–4 to each bundle, 2–5 mm. long, the connectives prolonged to form a short acute or bifid appendage. Ovary ovoid, the loculi with (1–)2–4 ovules; styles (3–)5–7 mm. long, fused for 2–3 mm. then spreading or recurved. Fruit green or crimson, then brown, broadly ellipsoid or globose, 1·5–4·5 cm. long, 2–3·5 cm. across, with copious yellow fluid turning brown on exposure to the air, 1–2(–3)-seeded, finely warty, crowned with reflexed stigma-lobes. Seed ovoid, compressed, 1·5–2 cm. long, 1–1·5 cm. wide, with a thin testa marbled by lactiferous ducts; inner integument fibrous. Fig. 2.

UGANDA. Ankole District: Kalinzu Forest, Dec. 1931, *R. A. Gibson* in *F.D.* 389!; Masaka District: Namalala Forest, Aug. 1913, *Fyffe* 81!; Mengo District: Sisa–Kisubi, Sept. 1937, *Chandler* 1915!
TANZANIA. Bukoba District: Kiamawa, Sept.-Oct. 1935, *Gillman* 420!; Tanga District: Mlinga Mt., 7 Dec. 1940, *Greenway* 6068!; Morogoro District: Salaza Forest, 4·8 km. S. of Bunduki, 19 Mar. 1953, *Drummond & Hemsley* 1678!
DISTR. **U**2, 4; **T**1, 3, 6; W. Africa from Sierra Leone to Cameroun, S. Tomé, Gabon, Cabinda, Angola, Zaire, W. Zambia, and possibly in Madagascar (some of the supposedly endemic species are possibly conspecific with *S. globulifera*), also in tropical S. and C. America, S. Domingo, Jamaica, Dominica, Guadeloupe and Trinidad
HAB. Rain-forest, also fringing streams in swamp forest (often a dominant tree); 840–2550 m.

SYN. *S. globulifera* L. f. var. *africana* Vesque in DC., Monogr. Phan. 8: 230 (1893); Hiern, Cat. Afr. Pl. Welw. 1: 59 (1896); Z.A.E.: 560 (1914) & 14 (1922). Type: Angola, Golungo Alto, Alta Queta, above Undele, *Welwitsch* 1052 (LISU, holo., BM, K, iso.!)
S. globulifera L. f. var. *gabonensis* Vesque in DC., Monogr. Phan. 8: 231 (1893). Type: Gabon, near Libreville, *Jolly* 31 (P, holo.)
S. gabonensis (Vesque) Pierre in Bull. Soc. Linn. Paris, Bull. Mens. 155: 1228 (1896); C.F.A. 1: 130 (1937); F.P.N.A. 1: 626 (1948); T.T.C.L.: 244 (1949)
S. gabonensis var. *macrantha* Hutch. & Dalz. in F.W.T.A. 1: 235 (1927) & in K.B. 1928: 227 (1928); I.T.U.: 83 (1940), *nom. superfl.* Types: twelve syntypes cited including *Welwitsch* 1052

3. ALLANBLACKIA

Oliv. in J.L.S. 10: 42 (1867); Bamps in B.J.B.B. 39: 347–357, figs. 1, 2 (1969) & in Distr. Pl. Afr. 1, maps 12–21 (1969)

Stearodendron Engl. in N.B.G.B. 1: 43 (1895) & P.O.A. C: 275 (1895)

Trees with hollow longitudinally wrinkled branches. Leaves opposite, with secretory canals visible on the lower face. Flowers unisexual, in terminal racemes or panicles with secondary axes very reduced and sometimes accompanied by solitary or paired flowers in the axils of the upper leaves, or sometimes pseudo-axillary on short lateral branches. Sepals 5, imbricate, unequal. Petals 5, imbricate. Male flowers: stamens numerous, grouped in 5 very fleshy bundles opposite the petals; anthers subsessile in several rows on the inner face or on both faces of the bundle, longitudinally dehiscent; disc in the form of a star, with 5 ± clavate branches ending in 5 smooth glands, grooved superficially or pleated-laminated, alternating with the staminal bundles; ovary absent. Female flowers: 5 rudimentary staminal bundles alternating with 5 free disc glands; ovary conical, with 5 incompletely separated locules and parietal placentas; ovules 2–28 per locule, arranged in 2 rows; stigma peltate, sessile or subsessile. Fruit a berry, often very large. Seeds arillate.

A genus of 10 species restricted to tropical Africa. The seeds yield a vegetable butter and at Amani during the First World War were extensively used as a butter substitute in the manufacture of chocolate, etc.

Anthers arranged only on the internal face of the
    bundle (fig. 3/3, 5); fruits smaller, up to 20 cm.
    long:
      Pedicels 1–2 cm. long; leaf-blades narrowly oblong;
        fruit 5-ribbed   .   .   .   .   .   .   1. *A. kimbiliensis*
      Pedicels obsolete or up to 0·5(–1·2) cm. long; leaf-
        blades broadly oblong or often ± obovate;
        fruit not ribbed   .   .   .   .   2. *A. ulugurensis*
Anthers arranged on the 2 faces of the very fleshy
    bundle (fig. 3/2); pedicels (3·5–)6·5–8 cm. long;
    fruits very large, up to 34 cm. long .   .   .   3. *A. stuhlmannii*

1. **A. kimbiliensis** *Spirlet* in B.J.B.B. 29 : 357 (1959) & Guttif., Contr. Fl.
Congo : 86, t. 5 (1966); Bamps in B.J.B.B. 39 : 353, fig. 1/D (1969) & in F.C.B.,
Guttif. : 47, fig. 5/D (1970). Type : Zaire, Kimbili, *Michelson* 766 (BR, holo. !)

Tree 18–36 m. tall, with blackish brown glabrous branches; bark smooth,
slash exuding a yellow juice. Leaf-blades narrowly oblong, 8–22 cm. long,
2–4·5 cm. wide, acuminate at the apex, acute or obtuse at the base, coriaceous, glabrous; midrib prominent beneath, the lateral nerves numerous,
prominent on both faces, at right-angles to near margin and terminating in a
submarginal nerve; secretory canals generally superimposed on the lateral
nerves; petiole 1–2 cm. long, glabrous. Male flower: pedicels 1–1·6 cm. long,
glabrous; sepals green or greenish white, ± round, 0·8–1·3 cm. in diameter,
glabrous; petals white or cream tinged pink, ± round, ± 1·5–2·5 cm. in
diameter, glabrous; staminal bundles spade-shaped, 0·8–1 cm. long, 1–1·2 cm.
wide, the claw 1 cm. long, 2·5 mm. wide, the anthers placed only on the internal face; disc with glands smooth, rugose above, or only slightly grooved.
Female flower: pedicel 2 cm. long; sepals and petals ± as in male; staminal
bundles very reduced, ± 1·5 mm. long with a few subterminal anthers; disc
glands smooth; ovary 5-ribbed, glabrous with 2–4 ovules per locule; stigma
5-lobed. Fruits ovoid or subconical, 5-ribbed, 9–20 cm. long, 7–12 cm. in
diameter. Seeds 15–20, ellipsoid, 3–5 cm. long, 1·5–3·2 cm. in diameter. Fig.
3/5, 6.

UGANDA. Kigezi District : Ishasha Gorge, 10 Feb. 1945, *Greenway & Eggeling* 7104 ! &
    Impenetrable Forest, Apr. 1948, *Purseglove* 2675 B ! & Kayonza Forest near R.
    Ishasha, May 1940, *Cree* 369 ! & same locality, Oct. 1940, *Eggeling* 4191 !
DISTR. **U2**; Zaire
HAB. Rain-forest on steep rocky gorge slopes; 1500 m.

SYN. [*A. floribunda* sensu Dale & Eggeling, I.T.U., ed. 2 : 152 (1952), *non* Oliv.]

NOTE. The name *Allanblackia parviflora* A. Chev. appears on most herbarium sheets
    of this species of material collected in East Africa; it is a synonym of *A. floribunda* Oliv.

2. **A. ulugurensis** *Engl.* in E.J. 28 : 435 (1900); T.T.C.L. : 241 (1949);
Bamps in B.J.B.B. 39 : 354 (1969) & in Distr. Pl. Afr. 1, map 17 (1969). Types :
SE. Uluguru Mts., *Stuhlmann* 8773 (B, holo. †); Uluguru Mts., Lupanga Peak,
Schlieben 2958 (BR, neo. !, BM !, K !, P, isoneo.)

Evergreen tree or rarely shrubby, (3–)15–30 m. tall, with spreading open
branching; bole slightly buttressed, clear for 7·5 m.; bark brownish grey or
red-brown, finely reticulate; slash salmon-pink with yellow inner layer slowly
exuding a yellow latex. Leaf-blades oblong, elliptic or often slightly obovate-
oblong, 7·5–19·5 cm. long, 4–11 cm. wide, rounded or slightly emarginate at
the apex, broadly cuneate at the base, very leathery, the reticulate venation
very prominent on both surfaces when dry, margins inrolled; petiole
0·7–1·4 cm. long. Flowers clustered towards the ends of the branchlets,

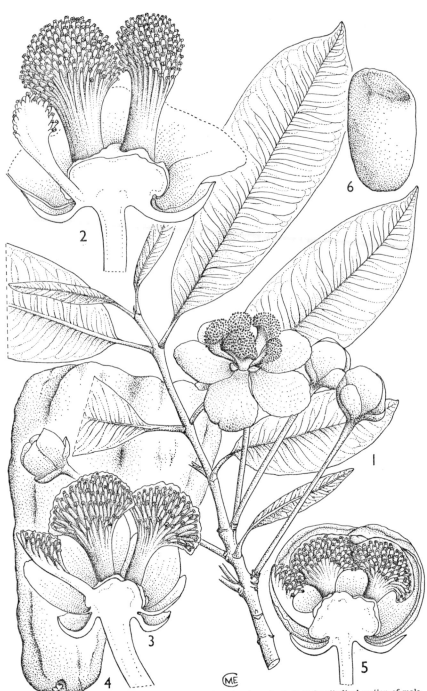

FIG. 3. *ALLANBLACKIA STUHLMANNII*—**1,** flowering branch, × ⅔; **2,** longitudinal section of male flower, × 2. *A. ULUGURENSIS*—**3,** longitudinal section of flower, × 2; **4,** fruit, × ⅔. *A. KIMBILIENSIS*—**5,** longitudinal section of flower, × 2; **6,** seed, × 1. 1, 2, from *Polhill* 4603; 3, 4, from *Polhill* 4605; 5, 6, from *Eggeling* 4191. Drawn by Mrs. M. E. Church.

axillary, fleshy; pedicels short, up to 0·5(–1·2) cm. long. Male flowers reddish
pink; sepals red-brown, elliptic or ± round, the inner 4–7·5 mm. long,
4–6·5 mm. wide, the outer ± round, up to 1·2 cm. long and wide; petals pink
or carmine or purplish, ± 1 cm. long and wide; staminal bundles clavate,
1–1·4 cm. long, widened above, ± 0·7–1·2 cm. wide, angled, the angle pointing
towards the centre of the flower. Fruits reddish pink but ochraceous when
dry, conical-oblong to conic, 10–13·5 cm. long, 6·5–8 cm. across, 2–few-seeded.
Fig. 3/3, 4.

TANZANIA. Morogoro District: Nguru Mts., saddle to NW. of Mkobwe, 29 Mar. 1953,
    Drummond & Hemsley 1900! & Uluguru Mts., Lupanga Peak track, 16 Aug. 1951,
    Greenway & Eggeling 8612! & Morningside, 14 Aug. 1959, *Mgaza* 291!; Iringa
    District: Ruaha valley, Ukwama, 26 Sept. 1958, *Ede* 5A!
DISTR. T6, 7; not known elsewhere
HAB. Rain-forest; 1000–2050 m.

3. **A. stuhlmannii** (*Engl.*) *Engl.* in E. & P. Pf., Nachtr. ii–iv Teil [1]: 249
(1897); Tropenpl. 3: 203–4, fig. (1899); V.E. 1(1): 296, t. 13, 19 (1910);
T.T.C.L.: 241 (1949); Bamps in B.J.B.B. 39: 356 (1969) & Distr. Pl. Afr. 1,
map 21 (1969). Type: Tanzania, E. Usambara Mts., Ngwelo [Nguelo], *Holst*
2296 (K, lecto.!)

Tree 12–36(?–45) m. tall, with bole clear for often about 9 m.; branches
drooping; bark dark grey, not or rarely flaking; slash with a clear exudate.
Leaf-blades oblong or elliptic-oblong, 5–19·5 cm. long, 1·2–7 cm. wide, shortly
acuminate at the apex, cuneate at the base, coriaceous or subcoriaceous, deep
green with midrib yellowish beneath; petiole 1–1·8 cm. long. Male and
female flowers large and fleshy, on different trees, both ♂ and ♀ solitary in the
axils but ♂ may be crowded at ends of shoots with short internodes and appear
terminally racemose if leaves have fallen; pedicels (3·5–)6·5–8 cm. long. Male
flowers: sepals pale yellowish, rounded ovate, the outer 4–8 mm. long, 8–9
mm. wide, the inner ± 2 cm. in diameter; petals cream, flushed scarlet at the
base or all scarlet, rounded or spathulate, ± 2·7–3·7 cm. long, 1·8–2·6 cm.
wide; staminal bundles crimson, often very unequal, thick and fleshy, 1·8–2
cm. long, 0·8–1(–1·7) cm. wide, 8 mm. thick, angled on inner face; anthers
yellow; disc-lobes green. Female flowers: similar to the male, the petals up
to 2·3 cm. long and wide; stamens reduced to a few free with filaments ± 4
mm. long; ovary ellipsoid or conic, 1·3–1·5 cm. tall, the stigma peltate,
sessile, 8–9 mm. wide, 3–4 mm. tall. Fruit brown or reddish brown, oblong,
tapering cylindric-oblong or subglobose, 16–34 cm. long, 15–17 cm. in diameter, weighing 5½–15 lbs. with (7–)12–28 seeds in each of 5 locules. Seeds
with 4 obtuse angles, 4 cm. long, 2–3 cm. wide, one angle with a fleshy aril;
testa crustaceous. Fig. 3/1, 2, p. 9.

TANZANIA. Lushoto District: Amani, 4 Feb. 1937, *Greenway* 4880 (♀)!, 4888 (♂)! &
    Amani, 14 Jan. 1950, *Verdcourt* 34! & 23 Jan. 1950, *Verdcourt* 56!; Morogoro District:
    Uluguru Mts., Mngazi [Mgasi] R. to Mdandsa R., Nov. 1898, *Goetze* 156!
DISTR. T3, 6, 7; not known elsewhere
HAB. Rain-forest; 540–1200 (–1600) m.

SYN. *Stearodendron stuhlmannii* Engl. in N.B.G.B. 1: 43 (1895) & P.O.A. C: 275 (1895)
    Allanblackia sacleuxii Hua in Bull. Mus. Nat. Hist. Nat. Paris 2: 155 (1896);
    Ann. Inst. Col. Marseille 5: 74 (1898). Type: Tanzania, Nguru Mts., *Sacleux*
    (P, holo.)

NOTE. The shape of the fruit varies considerably from tree to tree, *Verdcourt* 34 and
    56 cited above representing extremes; this variation has no taxonomic significance.

4. MAMMEA

L., Sp. Pl.: 512 (1753) & Gen. Pl., ed. 5: 228 (1754); Kosterm., The genera *Mammea* L. and *Ochrocarpos* Thou. (1956) & Comm. 72 of For. Res. Inst. Indonesia: 1–63 (1961);* Smith & Darwin in Journ. Arn. Arb. 55: 237 (1974)

Trees. Leaves opposite, petiolate, with numerous translucid dots and streaks in the areoles of the venation. Flowers polygamous, ♂ or ♀, the ♂ in axillary fascicles usually at already leafless nodes, the ♀ flowers solitary, axillary. Calyx entire in bud, dividing into 2(–3) segments during anthesis, ± persistent. Petals 4(–5–6 or more), imbricate, deciduous. Stamens numerous, free or joined at the base into a continuous ring. Ovary 2-locular, with 2 ovules per locule, or sometimes 4-locular, with 1 ovule in each locule due to the formation of false septa; placentation basal; style very short; stigma peltate, 2–4-lobed; ovary absent or much reduced in male flowers. Fruit a drupe, usually 1-locular by abortion but sometimes 4-celled, with 1–4 fibrous or woody 1-seeded pyrenes. Seeds large, with indistinct cotyledons usually surrounded by a mostly colourless edible pulp.

A genus of about 30 species in Asia, Africa and tropical America.

M. americana L. has been cultivated for its fruit the " Mamey apple ", e.g. Tanzania, Lushoto District, Sigi Plantation Sc. 6, 7 Mar. 1932, *Greenway* 2935! & Morogoro, Nov. 1955, *Semsei* 2385! also in Zanzibar at Migombani *fide* U.O.P.Z.: 339 (1949).

Vaughan 2310 (Zanzibar, Dongwe, 6 Jan. 1936), consisting of a leaf and a fruit, bears the data " large tree found only on the east coast—very rare ". The specimen appears to be *Mammea odorata* (Raf.) Kosterm. which is known from Java, Christmas I., Borneo, Moluccas, Ceram, Timor, New Guinea, Philippines, Solomon Is., Caroline Is., Admiralty Is., Marshall Is., Mariannas and other Pacific islands; also cultivated in Hawaii. A very full account may be found in Kostermans in Comm. 72 of the For. Res. Inst. Indonesia: 15–18, figs. 9, 10 (1961). The tree on Zanzibar was surely planted at some early date—if it was the result of a drift fruit then it would have to be considered native. The elliptic leaves (about 17 × 9 cm.) have a very close reticulation and are rounded; the fruit is curved-fusiform, 8·5 × 3 cm., pointed at both ends. More information is required about its status.

Leaf-blades rounded at the apex:
   Fruit ± globose; leaf reticulation with larger areoles   *M. americana* (see note above)

   Fruit fusiform; leaf reticulation with small areoles   *M. odorata* (see note above)

Leaf-blades acuminate at the apex:
   Pyrene-wall very woody, requiring a saw to cut it; fruit covered with small lenticel-like warts; ♂ flowers larger, the petals ± 1·5 cm. long (**U2**)   1. *M. africana*
   Pyrene-wall much thinner, with an anastomosing pattern of riblets which become more obvious when dry, easily broken with the fingers; fruit without lenticel-like warts; ♂ flowers smaller, the petals ± 1 cm. long (**T3**) . . .   2. *M. usambarensis*

1. **M. africana** *Sabine* in Trans. Hort. Soc. 5: 457 (1824); Staner in B.J.B.B. 13: 101 (1934); C.F.A. 1: 125 (1937) & 370 (1951); I.T.U., ed. 2: 153, fig. 32 (1952); Keay, F.W.T.A., ed. 2, 1: 293 (1954); Aubrév., Fl. For. Côte d'Ivoire, ed. 2, 2: 328, t. 251 (1959); Keay, Onochie & Stanfield, Nigerian Trees: 179 (1960); Bamps in F.C.B., Guttif.; 50, t. 8 (1970) & in B.J.B.B. 41: 427 (1971) & Distr. Pl. Afr. 3, map 60 (1971). Type: Sierra Leone, *Don* (where?)

* Kostermans actually treats *Ochrocarpos* as a synonym of *Mammea* in the literature citations but in his text states that he believes there is reason for keeping the two apart and Smith & Darwin have accepted this as a firm statement. *Ochrocarpos* would then be restricted to Madagascar and Kostermans also states that the frequent mention that the fruit is dehiscent in 2 Madagascan species is an error and that Perrier de la Bâthie misinterpreted fruits rupturing under pressure.

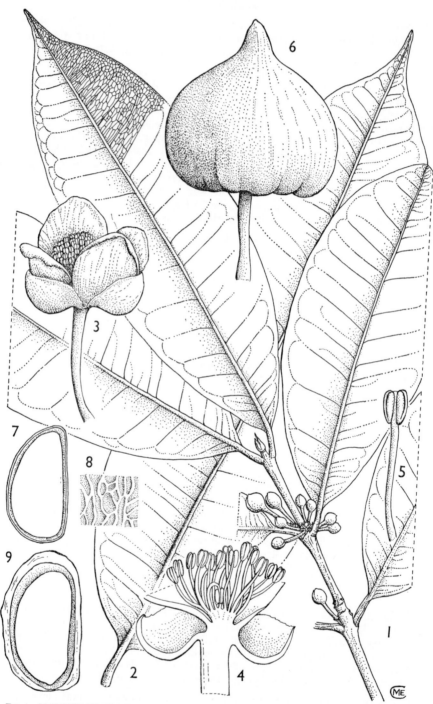

Fig. 4. *MAMMEA USAMBARENSIS*—**1,** flowering branchlet, × ⅔; **2,** leaf, × ⅔; **3,** male flower, × 3; **4,** longitudinal section of same, × 4; **5,** stamen, × 8; **6,** fruit, × ⅔; **7,** longitudinal section of pyrene, × ⅔; **8,** part of reticulate pyrene coat, × 2. *M. AFRICANA*—**9,** longitudinal section of pyrene, × ⅔. 1–8, from *Drummond & Hemsley* 2727; 9, from *Rowntree*. Drawn by Mrs. M. E. Church.

Tall tree (15–)30–45 m. tall, with spreading crown and greyish branches; bark often yellow or reddish brown with pale scales and with resinous yellow latex. Leaf-blades oblong, elliptic or oblong-elliptic, 9–35 cm. long, 3–13 cm. wide, acuminate at the apex, cuneate at the base, coriaceous, glabrous; mid-rib prominent beneath, the lateral nerves numerous, slightly prominent on both surfaces and reticulate; petiole 1–2·5 cm. long. Hermaphrodite flowers: pedicel 2–2·5(–3) cm. long; sepals red, ± round, concave, 1–1·5 cm. long and wide, glabrous; petals white or yellowish, elliptic to obovate, 1·5–2 cm. long, 0·6–1·2 cm. wide, glabrous; stamens fused into a continuous ring at the base; filaments 5–7 mm. long; anthers 2 mm. long; ovary ovoid; style 2 mm. long, with a stigma 4–5 mm. in diameter. Male flowers similar but smaller; pedicels 0·6–1·5 cm. long. Fruits orange with a yellowish pulp, subglobose or pyriform, (7·5–)10–18 cm. in diameter, the surface with numerous small brown spots; pyrenes brown, 1–4, ovoid, compressed, 4–5 cm. long, 3 cm. wide, 2 cm. thick, enclosed in a hard fibrous-woody wall. Fig. 4/9.

UGANDA. Bunyoro District: Budongo Forest, *Eggeling* 4979! & Budongo Forest, compartment N3, June 1970, *Kajaunbi* in *E.A.H.* 14381!
DISTR. U2; W. Africa from Sierra Leone to Cameroun, Principe, Gabon, Zaire and Angola
HAB. Rain-forest; 1050 m.

SYN. *Ochrocarpos africanus* Oliv. in F.T.A. 1: 169 (1868); F.W.T.A. 1: 235 (1927); Chalk et al., Twenty West African Timber Trees: 54, fig. 11 (1933); T.T.C.L.: 243 (1949), quoad descript. pro parte excl. spec. cit. Type: Principe, *Mann* 1119 (K, holo.!)
    Mammea ebboro Pierre in Bull. Soc. Linn. Paris, Bull. Mens. 154: 1223 (1896); Engl. in E.J. 40: fig. 3 on p. 565 (1908); V.E. 3(2): 507, fig. 230 (1921). Types: Gabon, near Libreville, *Klaine* 27 & 76 (P, syn.)
    Garcinia golaensis Hutch. & Dalz., F.W.T.A. 1: 237 (1927) & in K.B. 1928: 228 (1928). Type: Sierra Leone, Gola, *Unwin & Smythe* 67 (K, holo.!)

NOTE. It has been suggested that both specimens cited above come from the same tree; only one is known to foresters in the area.

2. **M. usambarensis** *Verdc.* in K.B. 31: 259, fig. 1 (1976). Type: Tanzania, W. Usambara Mts., W. Shagayu Forest, *Drummond & Hemsley* 2727 (K, holo.!, EA, iso.)

Tree to 24 m. tall and girth attaining over 6 m., with a clean straight bole and a rounded crown; bark brownish grey or reddish brown, smooth or obscurely flaking into fairly thick irregular flakes; slash red, slowly exuding a pale yellow gummy latex. Leaves drying yellow-green; blades narrowly to broadly oblong-elliptic, 13·5–29 cm. long, 6·4–9 cm. wide, acuminate or abnormally rounded at the apex, cuneate at the base, coriaceous; petiole 1·1–1·4 cm. long. Male flowers up to ± 6 per node; pedicels ± 1 cm. long; calyx pale green, splitting into 2 very concave sepals 7·5 mm. long, 6 mm. wide if not flattened out; petals 4, white, ovate, 1 cm. long, 7 mm. wide; stamens exceeding 100, the filaments cream, ± 4 mm. long; anthers deep yellow. Female flowers not seen. Fruits pale green or almost yellowish when mature, subglobose, 4–6·5 cm. long, 3·5–7 cm. wide, very distinctly acuminate-mammate at the apex, ± flattened below around junction of pedicel which is 2–3·3 cm. long; pyrenes brown, ± ½-globose, 3·7–4 cm. long, 3·7–4·5 cm. wide, 2–2·2 cm. thick, the wall thin, covered with ± obscure anastomosing ribs which become very evident and raised when dry. Fig. 4/1–8.

TANZANIA. Lushoto District: W. Shagayu Forest, 24 May 1953, *Drummond & Hemsley* 2727! & Shagayu Forest, May 1953, *Procter* 184! & Magamba Forest, 17 Sept. 1934, *Pitt-Schenkel* 393!
DISTR. T3; not known elsewhere
HAB. Evergreen forest; (1650–)1800–2100 m.

SYN. [*Ochrocarpos africanus* sensu Brenan, T.T.C.L.: 243 (1949) quoad *Pitt-Schenkel* 233 & 393, excl. descr. etc., *non* Oliv.]

[*Mammea africana* sensu Bamps in B.J.B.B. 41: 427 (1971) & in Distrib. Pl. Afric. 3, map 60 (1971), *non* Sabine]

NOTE. This has been misidentified with *Mammea africana*, but the fruit structure is quite different. In *M. africana* the pyrene needs cutting with a saw to get inside, but that of *M. usambarensis* can be broken with the fingers. It is said to be a dominant tree where it occurs, producing vast quantities of fruit.

5. GARCINIA

L., Sp. Pl.: 443 (1753) & Gen. Pl., ed. 5: 202 (1754); Pierre, Fl. Forest, Cochinchin. 1: Enum. Espèces du Genre *Garcina* I-XL, t. 54–83 (1882–83)

Rheedia L., Sp. Pl.: 1193 (1753) & Gen. Pl., ed. 5: 499 (1754)

Trees or shrubs, rarely subshrubs, secreting a yellow latex when cut. Leaves opposite or sometimes subopposite or whorled, petiolate, entire, coriaceous or chartaceous, the venation usually ± prominent, often with translucent glandular canals and brownish resin canals; petiole often with a ± prominent ligulate appendage. Flowers terminal or axillary, solitary or in few–many-flowered cymes, fascicles, racemes or panicles, fewer flowered in ♀ or ♂ plants. Sepals 4, decussate, or sometimes 5, quincuncial, or 3, free. Petals 4–5(–8), greenish white to yellow. Male flowers: androecium of varied structure, mostly composed of 4(–5) fascicles of numerous stamens, each with filaments free or partially or completely fused together, the fascicles usually free in African species but in some extra-African ones the stamens can be totally joined to form a cup-like androecium; anthers occasionally transversely septate; sometimes with a whorl of sterile stamen bundles ("fasciclodes") alternating with the stamen-bundles or forming a cushion in which the stamens are inserted, often interpreted as a fleshy cup-shaped 4–5-lobed or entire disc; ovary rudiment sometimes present. Female and ♂ flowers: 4(–5) stamen-bundles or staminode-bundles similar to ♂ flowers but smaller and with few members and sometimes with fasciclodes, free or fused together in a ring at the base of the ovary; ovary globose, 2–5(–12)-locular, each locule with 1 apical ovule; style usually absent; stigma ± sessile, broad, 2–5-lobed or entire, sometimes quite large and inflated, sticky. Fruit a 1–4-seeded ± fleshy smooth or verrucose glabrous or puberulous berry, with large seeds coated in pulpy tissue.

A large genus of over 200 species confined to the tropics. Some Madagascan and New World species were formerly referred to a genus *Rheedia* L., but this is not now considered to be distinct. There are 10 native species in the Flora area and 4 others have been sparingly cultivated and are mentioned in the key. *G. mangostana* L., the Mangosteen, very much esteemed for its fruit, has been grown in Tanzania, E. Usambaras, Sigi, and on Zanzibar I. at Kibweni and Kizimbani (see U.O.P.Z.: 270, fig. (1949)). The seeds are formed parthenogenetically since in nearly all areas where it is native the male trees are absent or extremely rare. *G. ferrea* Pierre, a native of Indochina, has been grown at Amani (Tanzania, Lushoto District, Amani, 24 June 1929, *Greenway* 1603! & 1 Feb. 1921, *Soleman in Herb. Amani* 6068! & 17 July 1971, *Ngonyani* 18!). *G. indica* DC., the kokum, the fruit of which is used in curries, has been grown in Zanzibar I. at Migombani and Mtoni. *G. xanthochymus* Hook. f. (*Xanthochymus pictorius* Roxb. *non G. pictoria* Roxb.), a native of India and one of the species from which the pigment gamboge is obtained, has also been cultivated at Amani, e.g. 7 Nov. 1932, *Greenway* 2941! and also in Zanzibar *fide* U.O.P.Z.: 272 (1949); there has been some confusion over the correct name for this species.*

G. wentzeliana Engl., described from Tanzania, Uhehe, and based on *Goetze* 440, is in fact a species of *Salacia* (T.T.C.L.: 243 (1949)).

* Brenan (T.T.C.L.: 242 (1949)) uses the name *G. tinctoria* (DC.) Dunn (1915), but he cites the origin of *Xanthochymus tinctorius* as being Roxb. in Pl. Coromandel where the name is spelt *pictorius*; De Candolle's use of a different name is illegitimate and there is in any case an earlier combination *G. tinctoria* (DC.) W. F. Wight (1909). Kurz also published the name *G. roxburghii* (1875), but Hooker's replacement name has priority by one year.

Stamens in ♂ flowers fused into a hemispherical or
cylindrical mass, the anthers ± subsessile in 3–4
ranks around the mass (cultivated) . . *G. indica*
(see note above)

Stamens in ♂ flowers not fused to make a single mass
but free or in bundles :
Staminal bundles 4–5 or stamens free (to p. 16) :
Stamen-filaments incompletely fused or free;
fasciclodes present; petiolar ligule prominent
(to p. 16) :
Staminal bundles 5 (rarely 4); stigma 5-lobed;
inflorescences terminal or axillary; seeds
large (sect. *Xanthochymus* (Roxb.) T.
Anders.) :
Ovary narrowed into a distinct style; stigma
with 5 very distinct spreading spathulate
divisions; leaves large, broadly oblong-
lanceolate, often exceeding 30 cm.; inflorescences axillary (cultivated) . . *G. xanthochymus*
(see note above)

Ovary not narrowed into a style; stigma
peltate, 5-lobed; leaves up to 20 cm.
long; inflorescences terminal . . 7. *G. volkensii*
Staminal bundles 4, the filaments united for
± ⅔ of their length or less or filaments
free; stigma (except *G. mangostana*)
2–3(–4)-lobed; inflorescences axillary in
native species :
Stamens not in bundles but free, inserted on a
fleshy cushion or annulus formed of fused
fasciclodes; leaves opposite or in whorls
of 3 or rarely 4 :
Female flowers ± 5 cm. across; male
flowers scarcely known, the seeds de-
veloping parthenogenetically; ovary
4–8-locular; stigma 5–8-lobed; row of
free stamens around the ovary; male
flowers in 3–9-flowered inflorescences
with stamens in a 4-lobed mass (cultivated for fruit; sect. *Garcinia* (sect.
Mangostana Choisy)) . . . *G. mangostana*
(see note above)

Female flowers much smaller (sect. *Tera-
centrum* Pierre) :
Leaves usually in whorls of 3, drying
pale green; petals 3–11 mm. long :
Bark not corky; petiole 4–10 mm.
long; leaf-margin plane or slightly undulate, not or only slightly
incrassate . . . . 1. *G. livingstonei*
Bark corky; petiole 2–4 mm. long;
leaf-margin strongly undulate,
markedly incrassate . . 2. *G. pachyclada*
Leaves opposite, drying reddish brown
beneath; petals 12–16 mm. long . 3. *G. semseii*
Stamens in 4 bundles, alternating with 4
fasciclodes; leaves opposite (sect.
Rheediopsis Pierre) :

Pedicels and sepals distinctly crimson; leaf-blades ± oblong, usually rounded or cordate at the base, but very variable and often cuneate; staminal bundles of 6–10 stamens; pedicels (1–)1·5–4·5 cm. long; petiole 2–4 mm. thick .   .   .   .   4. *G. smeathmannii*

Pedicels and sepals not crimson; leaf-blades oblong to oblong-lanceolate, mostly cuneate or sometimes ± rounded in *G. ovalifolia*; petiole 1–2 mm. thick:

Staminal bundles of 3(–4) stamens; pedicels 1·5–5 mm. long; main lateral nerves tending to curve upwards (**U**1)   .   .   .   5. *G. ovalifolia*

Staminal bundles of (5–)8–10 stamens; pedicels 0·7–1·5(–2·7) cm. long; main lateral nerves ± straight (**T**1, 6–8) .   .   .   .   6. *G. kingaensis*

Stamen-filaments completely fused, the anthers practically sessile; fasciclodes absent and ligule inconspicuous in native species:

Flowers large; rudimentary ovary well developed in ♂ flowers, ± cylindrical-obconic, much widened to the apex, not distinctly lobed; androecium 4-lobed (sect. *Kiras* Pierre) .   .   .   .   .   .   *G. ferrea*
(see note above)

Flowers smaller; rudimentary ovary not well developed; anthers oblong (sect. *Tagmanthera* Pierre):

Leaf-blades ± long-acuminate; petiole 0·5–1·5 cm. long; bark not corky (widespread) .   .   .   .   .   8. *G. buchananii*

Leaf-blades rounded, obtuse, acute or apiculate at the apex; petiole 2–6 mm. long; bark corky (**T** 1, 4, 7) .   .   .   9. *G. huillensis*

Staminal bundles 2, opposite the sepals (unnamed group close to sect. *Tagmanthera* Pierre):

Branchlets slender, narrowly 4-winged; leaf-blades up to 9×5 cm., acutely acuminate; petals 3–3·5 mm. long in ♂ flowers   .   10. *G. acutifolia*

Branchlets at first 4-angled, later rounded; leaf-blades up to 15×7 cm., obtuse or obtusely acuminate; petals ± 7 mm. long in ♂ flowers   11. *G. bifasciculata*

1. **G. livingstonei** *T. Anders.* in J.L.S. 9: 263 (1866); Oliv., F.T.A. 1: 165 (1868); Vesque in DC., Monogr. Phan. 8: 337 (1893); P.O.A. C: 275 (1895); T.T.C.L.: 241 (1949); Aubrév., Fl. For. Soud.-Guin.: 148, t. 24/3–4 (1950); Keay, F.W.T.A., ed. 2, 1: 294 (1954); Pardy in Rhodes. Agric. Journ. 53: 958, fig. (1956); Keay, Onochie & Stanfield, Nigerian Trees 1: 182 (1960); K.T.S.: 231 (1961); Robson in F.Z. 1: 400 (1961); Bamps in B.J.B.B. 39: 358 (1969) & in Distr. Pl. Afric. 1, map. 22 (1969); Killick & Robson in F.S.A. 22: 21 (1976). Type: plant cultivated at Calcutta, originally from Mozambique collected by Kirk (CAL, holo.)

Shrub or small tree (0·9–)3–18(–21) m. tall, usually of gnarled appearance, with dense mostly rounded crown, short twisted bole and long horizontal or

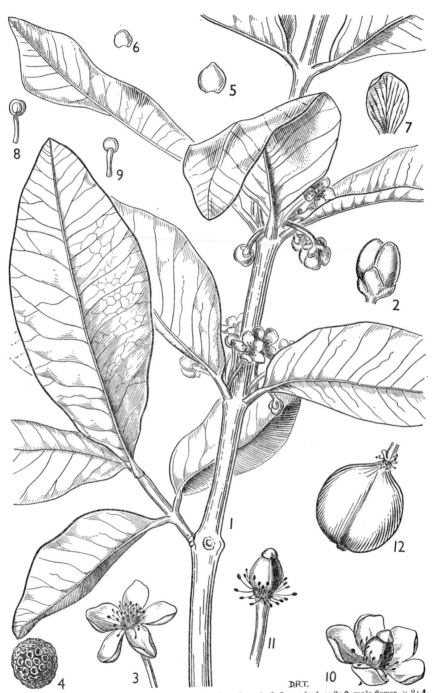

Fig. 5. *GARCINIA LIVINGSTONEI*—**1**, flowering branch, × 1; **2**, flower-bud, × 2; **3**, male flower, × 2; **4**, receptacle from above, × 4; **5, 6**, sepals, × 2; **7**, petal, × 2; **8, 9**, stamen, front and back views, × 4; **10**, hermaphrodite flower, × 2; **11**, same with sepals and petals removed, × 2; **12**, fruit, × 1. All from *van Someren* in *C.M.* 5099. Drawn by Miss D. R. Thompson.

pendulous branches; bark rough or ± smooth, usually slightly fissured, the slash mostly exuding a yellow or red resinous latex which may be almost lacking in specimens growing in rocky areas. Leaves mostly in whorls of 3 (rarely 4 or opposite); blades exceedingly variable in shape, lanceolate to oblanceolate to oblong or obovate or even ± round, 4–14(–17) cm. long, 1·5–11·5 cm. wide, emarginate or rounded to acute or apiculate at the apex, cuneate to rounded at the base, mostly very coriaceous, entire to somewhat crenate, pale to dark green, ± glaucous beneath; venation prominent and reticulate on both surfaces; secretory canals not visible; petiole 4–8 mm. long, channelled above, transversely wrinkled, with ligule prominent. Flowers sweet-scented and very attractive to insects, polygamous, in crowded fascicles of 5–15 or more on the older mostly leafless parts of branchlets but sometimes in the axils of older leaves; pedicels red, 0·4–2(–3·5) cm. long, varying in thickness. Sepals green, unequal, 4, in 2 opposite and decussate pairs, or 3, oblong or ± round, (1–)2–5·5 mm. long (1–)3–6 mm. wide, cucullate. Petals 5(–8), green or creamy white to pale yellow, with translucent colourless margins and ± translucent orange or reddish longitudinal glandular lines, obovate, elliptic or rounded, 3–6(–11) mm. long, 3–6·5(–11) mm. wide. Male flowers with numerous apparently free stamens inserted in a fleshy subglobose cushion 5 mm. wide, made up of the united fasciclodes; filaments white, 2–2·5 mm. long; anthers yellow or reddish. Hermaphrodite and ♀ flowers with fewer stamens or staminodes inserted on a fleshy ring below the ovary; ovary globose, 2(–3)-locular; stigma greenish white, fleshy, adnate, 2-lobed. Fruit yellow, red or orange, sometimes pinkish tinged, plum-like, ovoid, ellipsoid, obovoid or globose, 1–2·5(–4) cm. long, 1–3 cm. in diameter, sometimes compressed, edible. Seeds 1–2(–3), oblong-ellipsoid, cylindric or plano-ovoid, 1·5–2·1 cm. long, 8–11 mm. in diameter or ± 8 mm. thick and ± 1·1 cm. wide. Fig. 5.

UGANDA. Karamoja District: Moroto, 19 Feb. 1953, *Dawkins* 786! & Amudat, 22 Feb. 1953, *Dawkins* 794! & Lodoketeminit, 6 Nov. 1962, *Kerfoot* 4418!
KENYA. Machakos District: Katumani Experimental Farm, 5 Oct. 1958, *Napper* 843!; Masai District: Mara Masai Reserve, Telek R., 10 Sept. 1947, *Bally* 5287!; Kwale District: S. of Mrima Hill, 8 Sept. 1957, *Verdcourt* 1936!
TANZANIA. Tanga District: Sawa Creek, 16 Sept. 1957, *Faulkner* 2068! & 16 Nov. 1957, *Faulkner* 2093!; Kondoa District: Bubu valley, near Serya [Salia], 23 Feb. 1928, *B.D. Burtt* 1519! & 7 Jan. 1928, *B.D. Burtt* 1123!; ?Mbeya District: near Lake Rukwa, gorge of R. Mburu (?Mbowu), 18 Aug. 1935, *Michelmore* 1411!; Zanzibar I., Unguja Ukuu, 5 Dec. 1930, *Greenway* 2662!
DISTR. U1; K1, 4–7; T1–8; Z; Guinée to Cameroun, Somalia, Zambia, Rhodesia, Malawi, Mozambique, Angola, Botswana, South Africa (Transvaal and Natal), South West Africa (Caprivi Strip) and Ngwane
HAB. Woodland, thicket and grassland, nearly always riparian but also on rocky outcrops away from water and in open coastal forest of *Parkia*, *Trachylobium*, etc.; 0–1650 m.

SYN.   *G. angolensis* Vesque in DC., Monogr. Phan. 8: 335 (1893); R.E. Fries, Wiss. Ergebn. Schwed. Rhod.-Kongo-Exped. 1: 151 (1914); Staner in B.J.B.B. 13: 120 (1934); C.F.A. 1: 127 (1937). Types: Angola, R. Lifune, Libongo, *Welwitsch* 1047 (LISU, syn., BM!, K!, P, isosyn.) & Bumbo, near Quitibe de Cima, *Welwitsch* 1048 (LISU, syn., BM!, COI, P, isosyn.)
    G. pendula Engl. in E.J. 40: 557 (1908); T.T.C.L.: 242 (1949). Type: Tanzania, W. Usambara Mts., Rusotto to Mazinde [Masinde], *Busse* 359 (B, holo. †, BM, rough sketch!)
    G. ferrandii Chiov., Result. Sci. Miss. Stef.-Paoli, Coll. Bot. 1: 26, t. 21/A (1916) & Fl. Somala 2: 18, fig. 5 (1932). Types: Somalia, R. Juba, Biobahal, *Paoli* 847 & Lugh, *Paoli* 1012 (both FI, syn.)
    G. ferrandii Chiov. var. *affinis* Chiov., Result. Sci. Miss. Stef.-Paoli, Coll. Bot. 1: 27 (1916) & Fl. Somala 2: 19, fig. 6 (1932). Type: Somalia, Mahaddei, Uen, *Paoli* 1290 (FI, holo.)
    G. livingstonei T. Anders. var. *pallidinervia* Engl. in E.J. 55: 389 (1919). Types: Tanzania, Rungwe District, Kilambo, near Kajala on Ngubwisi stream, *Stolz* 1642 (B, syn. †, K, isosyn.!) & Lufirio [Lufiljo] R., *Stolz* 2237 (B, syn. †)
    G. pallidinervia (Engl.) Engl. in V.E. 3(2): 510 (1921) & in E. & P. Pf., ed. 2, 21: 216 (1925); T.T.C.L.: 242 (1949)

G. sp. aff. *livingstonei* T. Anders. sensu I.T.U., ed. 2: 153 (1952)

NOTE. *G. kilossana* Engl. in N.B.G.B. 2: 189 (1898), based on Tanzania, Kilosa, *Brosig* (B, holo. †, BM, rough sketch), may belong here, but we have seen no authentic material. The BM sketch bears the data " type " and " Sachsenwald bei Dar es Salaam, *Stuhlmann* ". Brosig gives the vernacular name " mkowe ", but I have seen no specimen bearing this name from the area. Judging by the description given by Engler the type was sterile.

We have seen no authentic material of *G. pendula* and in fact no material from the W. Usambaras of any species resembling *G. livingstonei*. Robson formerly determined several sheets as *G. pendula* but Bamps has cited all as *G. livingstonei*. For the time being we have agreed to look on *G. livingstonei* as an exceptionally variable species.

We have also seen no material of *G. bussei* Engl. in E. & P. Pf., ed. 2, 21: 216 (1925); type: Tanzania, Tunduru District, Rovuma R., Chamba [Kwa-Schamba], *Busse* 1018 (B, holo. †); Exell made a sketch of the type (BM) and notes " leaf velutinous below, no flowers, fruit in alcohol not seen ". It seems probable that it belongs to another genus.

2. **G. pachyclada** N. *Robson* in Bol. Soc. Brot., sér. 2, 32: 170 (1958) & in F. Z. 1: 401 (1961); Bamps in B.J.B.B. 39: 361 (1969), in Distr. Pl. Afr. 1, map 23 (1969) & in F.C.B., Guttif.: 54 (1970). Type: Zambia, Mbala District, Chambezi R., *Michelmore* 511 (K, holo.!)

Shrubby tree 2·7–5 m. high, with the habit of *Uapaca kirkiana*, glabrous; branches thick, with a rough greyish and corky bark, yellowish and transversely plicate on the young shoots; latex yellow, sticky. Leaves in whorls of 3, subsessile or shortly petiolate; blades obovate to elliptic, 6–27·5 cm. long, 3·5–11 cm. wide, rounded to obtuse and often mucronulate at the apex, rounded to cuneate at the base, bright green and concolorous, the margins undulate and incrassate; venation prominent and reticulate on both surfaces; secretory canals not visible; petiole 2–4 mm. long, channelled above, transversely wrinkled; ligule prominent. Flowers unisexual, clustered on cushions in axils or above leaf-scars on the mostly leafless parts of branchlets; pedicels 5–22 mm. long. Male flowers 15–20 mm. in diameter; sepals 5, unequal, the 2 outer suborbicular, cucullate, 3 mm. long, 4 mm. wide, the 3 inner ovate to suborbicular, 7–10 mm. long, 6–8 mm. wide; petals 5, white or pale cream-yellow with translucent and reddish longitudinal glandular lines, similar to the inner sepals; stamens numerous, free, inserted in a fleshy cushion; filaments 2 mm. long; anthers ovoid; ovary vestige absent. Female flowers with fewer free staminodes inserted in a fleshy cushion; ovary globose or broadly ovoid, 2–3-locular; stigma sessile, fleshy, 2–3-lobed. Berry ovoid-globose, smooth, 1–3-seeded.

TANZANIA. Ufipa District: Kasanga area, Ngolotwa, 12 Dec. 1948, *Glover* in *Bredo* 6406!
DISTR. T4; Zaire, Zambia and Mozambique
HAB. Deciduous woodland and shrubby grassland; up to 1500 m.

3. **G. semseii** *Verdc.* in K.B. 31: 262, fig. 2 (1976). Type: Tanzania, Nguru Mts., Manyangu Forest, *Semsei* 1400 (K, holo.!, EA, iso.)

Tree 3–15 m.; branchlets slightly wrinkled and nodes rough with old petiole-bases. Leaves opposite; blades usually drying olive-brown above, reddish brown beneath, elliptic, (12–)19–25 cm. long, (4·5–)7–16·5 cm. wide, rounded or obscurely subacute at the apex, cuneate at the base, slightly shining above, the margins slightly crinkly, coriaceous; lateral nerves 24–35 pairs, with many additional less pronounced intermediate ones; petiole 1–2 cm. long. Flowers 1–2(or ? more) at leafless nodes; pedicels 1·8–3·1 cm. long; 3 outer sepals round to elliptic, 6–7 mm. long, 4–6·5 mm. wide; 2 inner sepals round, 1·1 cm. long, 1 cm. wide; petals pale yellow, obovate with thin edges, 1·2–1·6 cm. long, 0·9–1·3 cm. wide. Male flowers: stamens very

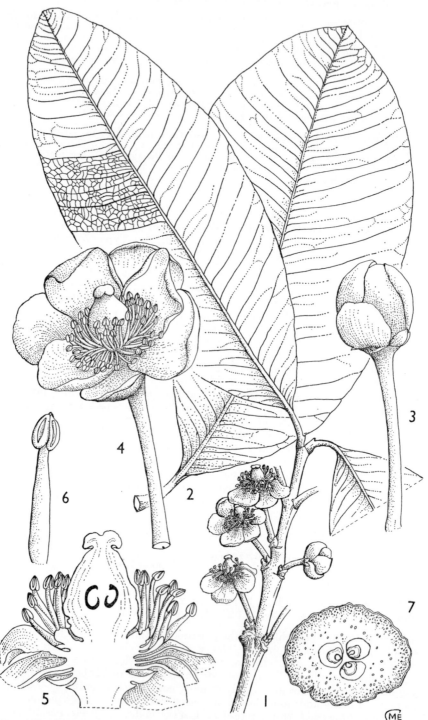

FIG. 6. *GARCINIA SEMSEII*—1, flowering branch, × ⅔; 2, leaf, × ⅔; 3, bud, × 2; 4, ⚥ flower, × 2; 5, longitudinal section of same, × 4; 6, stamen, × 8; 7, transverse section of ovary, × 4. All from *Semsei* 1400.

numerous, free on an elevated annulus 7–8 mm. in diameter; filaments 5–6 mm. long; anthers 0·8 mm. long. Hermaphrodite flowers: stamens numerous; filaments 4 mm. long; ovary resting on a very short thick gynophore slightly narrower than the ovary, 3 mm. tall, rugose and pitted; ovary conic, 6 mm. long, the base ± crenulate, 3-locular; stigma peltate, roundedly 3-lobed, 4 mm. wide. Fruit subglobose, 5 cm. long, 4 cm. across, roughly warted. Seeds 3, ellipsoid, 3·5 cm. long, 2·5 cm. wide. Fig. 6.

TANZANIA. Morogoro District: Nguru Mts., Turiani, Manyangu Forest, Nov. 1955, *Semsei* 1400!; Kimboza Forest Reserve, 27 Sept. 1971, *Pócs, Harris & Mwanjabe* 6466/A! & *Harris & Mwasumbe* in *DSM* 2428!
DISTR. **T6**; not known elsewhere
HAB. Rain-forest; 210–1800 m.

NOTE. This bears a remarkably close resemblance to the New World *G. megaphylla* Verdc. (*Rheedia macrophylla* Planch. & Triana) in general facies, some sheets seeming almost identical, but close examination shows a great difference in the structure of the stigma. The floral structure is exactly that of *G. livingstonei* of which it is a high forest analogue. I do not believe it is the species described as *G. pendula* Engl. since the flowers are very much larger.

4. **G. smeathmannii** (*Planch. & Triana*) *Oliv.* in F.T.A. 1: 168 (1868); Vesque in DC., Monogr. Phan. 8: 334 (1893); Keay, F.W.T.A., ed. 2, 1: 295 (1954); Robson in F.Z. 1: 398, t. 78 (1961); Bamps, F.C.B., Guttif.: 60 (1970). Type: Sierra Leone, *Smeathmann* (MPU, holo., BM, iso.!)

Tree (2·4–)4·5–21(–30) m. tall, the branches brownish grey, ridged longitudinally in the young state, glabrous; bark grey or yellowish to reddish brown, smooth with a yellow or red exudate. Leaves opposite; blade oblong or ovate to elliptic or obovate, 8–35 cm. long, 2·5–15 cm. wide, obtuse, acuminate or sometimes rounded at the apex, cuneate to rounded or subcordate at the base, coriaceous, glabrous; venation prominent on both surfaces, with 15–20(–25) pairs of lateral nerves; secretory canals ± parallel to the midrib, scarcely visible; petiole (0·8–)1–3 cm. long, 2–4 mm. thick, channelled and almost winged above, with a ligule at the base, transversely grooved, glabrous. Flowers in 5–30-flowered axillary fascicles inserted on little cushions mostly at the nodes on the old wood; pedicels crimson (1–)1·5–4·5 cm. long, glabrous. Male flowers: sepals crimson, 4, unequal, ovate to ± round, glabrous, the 2 outer 2–6 mm. long and wide, the 2 inner 3–7 mm. long and wide; petals creamy white, often tinged pink and with radiating translucent glandular lines, 4, obovate, 5–10 mm. long, 2–6 mm. wide, fleshy, glabrous; fasciclodes forming a 4-lobed "disc" alternating with the 4 bundles of stamens, which are formed of 6–10 stamens joined for ⅔ of their length and 3–4 mm. long with ovoid anthers. Female flowers similar to the male but staminal bundles reduced, of 1–4 stamens or staminodes alternating with small "discal glands"; ovary 2(–4)-locular, with a 2(–4)-lobed sessile stigma. Fruits purplish green turning yellow, subglobose or ± 2(–4)-lobed, (1–)2–2·5 cm. in diameter, smooth, glabrous. Seeds 1–2(–4), subglobose or plano-convex, 1–1·7 cm. long and 1–1·5 cm. wide.

TANZANIA. Buha District: 48 km. S. of Kibondo, Mukugwa [Mukugwe] R., July 1951, *Eggeling* 6214!; Rungwe District: Mwalesi R., 1 Mar. 1933, *R. M. Davies* 873!; Njombe District: upper Ruhudje R., Lupembe area N. of river, 18 Sept. 1931, *Schlieben* 1192!
DISTR. **T**1, 4, 7; Guinée to Cameroun and Fernando Po, Central African Empire, Gabon, Zaire, Malawi, Zambia and Angola
HAB. Riverine forest and bushland; 1000–1600 m.

SYN. *Rheedia smeathmannii* Planch. & Triana in Ann. Sci. Nat., sér. 4, 14: 312 (1860) *Garcinia polyantha* Oliv. in F.T.A. 1: 166 (1868); Staner in B.J.B.B. 13: 124 (1934), pro parte; C.F.A. 1: 127 (1937); T.T.C.L.: 242 (1949). Types: Nigeria, Nun R., *Mann* 488 (K, syn.!) & Lagos or Principe, *Barter* 1839 (K, syn.!) & Fernando Po, *Mann* 583 (K, syn.!)

G. chevalieri R.E. Fries, Wiss. Ergebn. Schwed. Rhod.-Kongo-Exped. 1: 151 (1914). Type: Zambia, Bwana Mkubwa, *Fries* 435 (UPS, holo. !)

G. mbulwe Engl. in E.J. 55: 389 (1919). Type: Tanzania, Rungwe District, Kiwira [Kibila], *Stolz* 933 (B, holo. †, BM, sketch !, K, iso. !)

G. stolzii Engl. in E.J. 55: 391 (1919). Type: Tanzania, Rungwe District, Kiwira [Kibila], *Stolz* 1589 (B, holo. †, BM, EA, K, iso. !)

5. G. ovalifolia *Oliv.* in F.T.A. 1: 166 (1868); Staner in B.J.B.B. 13: 129 (1934); Keay, F.W.T.A., ed. 2, 1: 295 (1954); Bamps in F.C.B., Guttif.: 61 (1970). Type: Nigeria, Nupe, *Barter* 807 (K, holo. !)

Shrub or tree 2·4–25 m. tall; young branches greyish, longitudinally grooved, glabrous; bark yielding yellow latex when cut. Leaves opposite; blades elliptic, obovate or oblanceolate, 4–15 cm. long, 1–6 cm. wide, obtuse to long-acuminate at the apex, cuneate to subrounded at the base, coriaceous, glabrous; venation prominent on both surfaces with numerous lateral nerves; secretory canals parallel to the midrib but little visible; petiole 0·4–1 cm. long, channelled above, ridged transversely, glabrous, with a ligulate appendage towards the base. Inflorescences fasciculate, axillary, several–many-flowered, inserted on little cushions at the nodes on the old wood; pedicels 1·5–5 mm. long, accrescent, glabrous. Male flowers: sepals 4, unequal, ovate to ± round, the 2 outside 1–1·5 mm. long and wide, the 2 inside 2–2·5 mm. long and wide, all glabrous; petals 4, obovate, 3–4 mm. long, 2–2·5 mm. wide, glabrous; fasciclodes forming a 4-lobed intrastaminal " disc ", alternating with the 4 staminal bundles, which are 2·5–3·5 mm. long of 3–4 stamens joined to each other in their lower ⅓; anthers ovoid. Hermaphrodite flowers resembling the male; stamens 4, free, alternating with the " disc " glands; ovary 2-locular; stigma subsessile, 2-lobed. Fruits ellipsoid, 1·5–2·5 cm. long, 1–2 cm. in diameter, with a smooth surface, glabrous. Seeds 1–2, plano-convex, 1·2–1·5 cm. long, 0·6–1 cm. wide.

UGANDA. W. Nile District: Amua, *Eggeling* 910 !

DISTR. U1; Mali and Guinée to Nigeria, Cameroun, Gabon, Central African Empire, Zaire, Sudan, Ethiopia and Angola

HAB. ?Streamside forest; 900 m.

SYN.   *G. pynaertii* De Wild. in Ann. Mus. Congo, Bot., sér 5, 2: 312 (1908). Types: Zaire, Eala, *Pynaert* 491, 935, 1484, 1765 & near Mogandjo, *Laurent* 2611 (all BR, syn. !)

      G. edeensis Engl. in E.J. 40: 560, fig. 1/J–K (1908) & V.E. 3(2): fig. 231/J–K (1921). Type: Cameroun, Edea, *Winkler* 892 (B, holo. †)

      G. claessensii De Wild. in F.R. 13: 373 (1914). Type: Zaire, Katako-Kombe, *Claessens* 369 (BR, holo. !)

      [*G. buchananii* sensu Dale & Eggeling, I.T.U., ed. 2: 153 (1952), pro parte, *non* Bak.]

NOTE. Bamps gives *G. edeensis* as figured in E.J. 40, fig. 1/E–H but on the plate it is fig. 1/J–K and this seems to fit the description; on the plate E–H are *G. dinklagei* Engl. Pencilled comments on the Kew copy of vol. 40 indicate the same idea as given by Bamps.

6. G. kingaensis *Engl.* in E.J. 30: 356 (1901); Bamps in B.J.B.B. 39: 371 (1969) & in Distr. Pl. Afr. 1, map 29 (1969). Type: Tanzania, Njombe District, Ukinga [Kinga] Mts., *Goetze* 1208 (B, holo. †, BR !, P, iso.)

Shrub or large tree 3–15 m. tall, glabrous; branches grooved, flattened or 4-angled when young, eventually cylindric; bark grey or brown, somewhat rough with fine reticulation; slash deep red with very fine drops of yellow latex. Leaves opposite; blades elliptic to elliptic-lanceolate or elliptic-oblong, 6–16 cm. long, 3–6 cm. wide, acute, obtuse or rarely shortly acuminate or apiculate at the apex, cuneate at the base, coriaceous, bluish to sage-green above, paler beneath, the venation prominent on both surfaces;

lateral nerves 30–60 pairs, leaving the midrib almost at right-angles; longitudinal glandular canals usually visible beneath; petiole 0·6–1·6 cm. long, channelled above, longitudinally grooved and transversely wrinkled, the ligule prominent. Flowers dioecious or ? polygamous, solitary or in fascicles of 2–7 in the axils of the older leaves; pedicels 0·7–1·5(–2·7) cm. long. Sepals 4, pale green, decussate, unequal, oblong to round, 1·5–5 mm. long, and wide, obtuse. Petals 4, greenish or creamy white or yellowish with longitudinal translucent glandular lines, ± round, 4–6(–10) mm. long, 5–6 mm. wide. Male flowers with 4 spongy fasciclodes united in the centre of the flower, alternating with 4 staminal bundles, each made up of (5–)8–10 stamens with the filaments connate for up to ⅔ of their length. Female flowers with small denticulate fasciclodes alternating with bundles of (2–)3–5 staminodes; ovary globose to ovoid, 2-locular; stigma fleshy, ± 2-lobed. Fruits orange-yellow when mature, globose, (1–)2·5–3·7 cm. long, 2·2–3·7 cm. wide, 1·7–2·8 cm. thick. Seeds 1–2, ovoid, (1–)2·6 cm. long, 2·1 cm. wide, 1·6 cm. thick, flattened on one side, the seed coat dark brown with some pale anastomosing lines.

TANZANIA. Morogoro District: Uluguru Mts., 17 Nov. 1934, *E. M. Bruce* 149!; Mbeya Range, 28 Nov. 1961, *Kerfoot* 3146! & Mbeya Peak Forest Reserve, 1 Sept. 1962, *Kerfoot* 4209!; Songea District: Liwiri [Luwiri]-Kiteza Forest Reserve, 16 Oct. 1956, *Semsei* 2522!
DISTR. T1, 6–8; Mozambique, Malawi, Zambia and Rhodesia
HAB. Evergreen forest, forest edges and upland grassland; 1200–2300 m.

SYN. [*G. polyantha* sensu Staner in B.J.B.B. 13: 124 (1934), pro parte quoad *Goetze* 1208; T.T.C.L. 2: 242 (1949), pro parte quoad *Pitt-Schenkel* 946, *non* Oliv.]
G. mlanjiensis Dunkley in K.B. 1937: 467 (1938); Robson in F.Z. 1: 400 (1961); F.F.N.R.: 225 (1962). Type: Malawi, Mt. Mulanje [Mlanje], *Burtt Davy* 22045 (K, holo.!)

7. **G. volkensii** *Engl.*, P.O.A. C: 275 (1895); T.T.C.L.: 243 (1949); Robson in F.Z. 1: 397 (1961); K.T.S.: 233, fig. 45 (1961); F.F.N.R.: 255 (1962); Bamps in F.C.B., Guttif.: 55 (1970) & in Distrib. Pl. Afr. 2, map 30 (1970) & in B.J.B.B. 40: 281 (1970). Type: Tanzania, Kilimanjaro, Useri, *Volkens* 1996 (B, holo. †, BR, iso.!)

Much-branched glabrous evergreen tree or shrub (2–)4–20 m. tall; branches stiff, grooved or ± winged, flattened, ± 3–4-angled or rounded; bark grey-brown, smooth, yielding a milky juice or yellow latex when cut. Leaves opposite or rarely in whorls of 3–4; blades very variable, lanceolate or oblanceolate to broadly ovate or obovate, (1·2–)4–20 cm. long, (0·9–)1·5–8 cm. wide, rarely exceeding 11 × 4·5 cm., acute to rounded at the apex but apiculate, cuneate to rounded at the base, mostly dark green above and yellowish or pale green beneath, mostly distinctly coriaceous, the margin often thickened; venation mostly prominent on both faces when dry, the surface often rugulose, with translucent and opaque canals usually visible; petiole 0·3–1·8 cm. long, with prominent ligule 2–4 mm. long. Flowers dioecious, in terminal lax ± condensed 1–many-flowered cymose inflorescences; peduncle 0–4 cm. long; bracts scale-like, keeled; pedicels 0–2 mm. long. Sepals 5, reddish, ± unequal, rounded to triangular-ovate, 1–2·5 mm. long and wide, glabrous. Petals 5, cream to greenish white, sometimes tinged with pink, with yellowish linear glands radiating from the base, fleshy, round to broadly obovate, 4–9 mm. long, 3·5–7 mm. wide, glabrous. Male flowers with 5 yellow spongy honeycombed fasciclodes uniting to form a star in the centre of the flower, alternating with 5 greenish cream staminal-bundles, the filaments joined for most of their length, each bundle 3·5 mm. long, 0·7 mm. wide, with 5–9 red, brown or mustard-coloured anthers. Female flowers with small fasciclodes alternating with sterile staminal-bundles; ovary green or yellowish, 2–4-locular; stigma white, peltate, 5-lobed. Fruit green turning yellow, reddish brown or orange, globose, ovoid or 2–4-lobed, 1–3(–5) cm. in diameter, smooth and glabrous. Seeds red, 1–4, ovoid, 1–2·2(–4·5) cm. long, 0·7–2(–3) cm. wide and 0·6–1·4 cm. thick, often much compressed.

Kenya. Naivasha District: S. Kinangop, Sasamua Dam pipe-line road, 11 Dec. 1960, *Verdcourt, Polhill & Lucas* 3035 !; Kiambu District: Gatamaiyo Forest [Katamayo], near to Kerita Forest Station, 16 Nov. 1958, *Verdcourt* 2306 !; Teita Hills, 8 km. NNE. of Ngerenyi, 15 Sept. 1953, *Drummond & Hemsley* 4344 !

Tanzania. Moshi District: Barankata area, Nov. 1960, *Steele* 113 !; Lushoto District: Shagayu [Shaga] Forest, near Sunga West, 17 May 1953, *Drummond & Hemsley* 2587 !; Morogoro District: Uluguru Mts., Lukwangule Plateau, 30 Jan. 1935, *E. M. Bruce* 705 !

Distr. **K**3, 4, 6, 7; **T**2, 3, 5–7; Zaire, Rwanda, Burundi, Mozambique, Malawi and Zambia

Hab. Evergreen forest; (30–)960–2400 m.

Syn. *G. usambarensis* Engl. in E.J. 40: 561 (1908); T.T.C.L.: 242 (1949). Type: Tanzania, Lushoto District, near Derema [Nderema], Ngambo, *Scheffler* 190 (B, holo. †)
　　　G. albersii Engl. in E.J. 40: 512 (1908); T.T.C.L.: 243 (1949). Type: Tanzania, W. Usambara Mts., Kwai, *Albers* 349 (B, holo. †, EA, iso. !)
　　　G. bangweolensis R.E. Fries in Wiss. Ergebn. Schwed. Rhod.-Kongo-Exped. 1: 152, t. 11 (1914). Type: Zambia, Lake Bangweulu, N. of Kasomo, *Fries* 708 (UPS, holo. !)

Note. A specimen from Mafia I., Kikuni, 13 Aug. 1937, *Greenway* 5084 !, which is the origin of the 30 m. altitude given above, is from an abnormally low site but several species reach low levels on this island due to causes as yet unknown.

8. **G. buchananii** *Bak.* in K.B. 1894: 354 (1894); Fries, Wiss. Ergebn. Schwed. Rhod.-Kongo-Exped. 1: 153 (1914); Staner in B.J.B.B. 13: 133 (1934); T.T.C.L.: 242 (1949); I.T.U., ed. 2: 153 (1952); Bamps in B.J.B.B. 39: 367 (1969) & in Distr. Pl. Afr. 1, map 26 (1969) & in F.C.B., Guttif.: 66 (1970). Type: Malawi, without precise locality, *Buchanan* 183 (K, holo. !)

Tree or shrub 1·5–15(–25) m. tall, with dark grey or brownish non-corky bark yielding yellow latex when cut; young branchlets red, angular. Leaves opposite; blades elliptic to ovate, (3·5–)7–16 cm. long, (2–)2·5–7·5 cm. wide, ± long-acuminate at the apex, cuneate, obtuse or sometimes ± rounded at the base, usually ± coriaceous, glabrous; venation prominent on both faces with numerous lateral nerves; secretory canals continuous and opaque, ± parallel to the midrib, visible beneath; petioles red, (0·5–)1–1·5 cm. long. Corolla white or creamy yellow to orange. Male flowers in short axillary or terminal cymes, often raceme-like; pedicels 3–5 mm. long; sepals 4, ± round, unequal, the outer 2–2·5(–5) mm. long and wide, the inner 3·5–4(–10) mm. long and wide; petals 4, obovate, 8–9(–12) mm. long, 4–5(–10) mm. wide, glabrous; fasciclodes joined to form an obconic " disc " 2 mm. in diameter; staminal bundles 4, of 5–6 sessile stamens each; anthers not septate. Female flowers usually solitary, axillary or terminal; pedicel 3–6 mm. long; sepals and petals resembling those of the male flowers; stamens absent or sometimes a few present, free and sterile; ovary 4-locular; stigma orange, sessile, 4-lobed. Fruit orange, yellow or red, subglobose, 2–2·5 cm. in diameter, with a smooth surface, glabrous, edible. Seeds 1–4, ellipsoid, 0·7–1·5 cm. long, 6–8·5 mm. wide, 3–5 mm. thick.

Uganda. Mbale District: Bugisu, Bulisegi, Aug. 1936, *Tothill* 2603 ! & Budadiri, Aug. 1936, *Tothill* 2604 !; Masaka District: Sese Is., Bugala I., Sozi Point, Nov. 1931, *Eggeling* 80 ! & Lake Nabugabo, July 1937, *Chandler* 1747 !

Kenya. ?Central Kavirondo District: Yala R., 16 June 1953, *G. R. Williams & Piers* 589 !; S. Kavirondo District: Watende, 14 Apr. 1955, *Argyle* 119 !; Kwale District: Buda-Mafisini Forest, some 4·8–6·4 km. inland from Msambweni, 21 Jan. 1964, *Verdcourt* 3955c !

Tanzania. Mwanza District: Ukerewe I., 25 Sept. 1931, *Conrads* 5984 ! & May 1930, *Conrads* 5130 !; Mpanda District: Mahali Mts., Kasangazi R., *Jefford, Juniper & Mgaza* 241 !; Lindi District: Rondo Plateau, Mchinjiri, Feb. 1952, *Semsei* 638 !

DISTR.   U3, 4; **K**2, 5–7; **T**1, 3–8; Zaire, Rwanda, Burundi, Sudan, Mozambique, Malawi, Zambia and Rhodesia

HAB.   Evergreen forest, often riverine, thickets and densely wooded grassland, also in coastal forest on pure sand; 60–1800 m.

SYN.   [*G. huillensis* sensu Milne-Redh. in Mem. N.Y. Bot. Gard. 8: 222 (1953); Robson in F.Z. 1: 402 (1961), pro parte; K.T.S.: 231 (1961); F.F.N.R.: 255 (1962), pro parte, *non* Oliv.]

NOTE.   See note at the end of *G. huillensis*.

9.   **G. huillensis** *Oliv.* in F.T.A. 1: 167 (1868); Vesque in DC., Monogr. Phan. 8: 353 (1893); Hiern, Cat. Afr. Pl. Welw. 1: 61 (1896); Staner in B.J.B.B. 13: 134 (1934), pro parte; C.F.A. 1: 128 (1937); T.T.C.L.: 242 (1949); Robson in F.Z. 1: 402 (1961), pro parte; F.F.N.R.: 255 (1962), pro parte; Bamps in B.J.B.B. 39: 368 (1969) & in Distr. Pl. Afr. 1, map 27 (1969) & F.C.B., Guttif.: 67 (1970). Type: Angola, Huila, Morro de Lopollo, *Welwitsch* 1051 (LISU, holo., BM!, COI, K!, P, iso.)

Shrub or small tree 2–5(–8) m. tall, with angular young branchlets; bark grey to ochraceous, corky. Leaves opposite or subopposite; blades obovate to elliptic, (2·5–)6–10 cm. long, (0·9–)1·5–6 cm. wide, rounded, obtuse or pointed and sometimes apiculate at the apex, often mucronulate, cuneate to attenuate-decurrent at the base, coriaceous or almost so, glabrous; nerves prominent on both faces, the lateral ones numerous, crossed by opaque secretory canals scarcely visible beneath; petiole 2–6 mm. long, channelled above. Corolla white or yellow. Male flowers arranged in short axillary cymes, often raceme-like; pedicels 2–5 mm. long; sepals 4, ± round, unequal, the outer 2·5–3·5 mm. long and wide, the inner 5–6 mm. long and wide; petals 4, obovate, 7–9 mm. long, 4–5 mm. wide, glabrous; staminal bundles 4, 4 mm. long, made up of 5–8 sessile stamens; anthers not septate; fasciclodes forming an obconic " disc " 2 mm. long. Female flowers usually solitary, axillary or terminal; pedicel 2 mm. long; sepals and petals as in the male flower; stamens absent; ovary 4-locular; stigma sessile, 4-lobed. Fruits red or orange, sub-globose, 1·5–3 cm. in diameter with smooth surface, glabrous. Seeds 1–3(–4), ellipsoid, slightly reniform, 1·2–1·5 cm. long, 0·7–1 cm. wide, 6–8 mm. thick, with a thin brittle veined integument.

TANZANIA.   Mwanza District: 24 km. S. of Geita Gold Mine, 13 Apr. 1937, *B. D. Burtt* 6514!; Buha District: Gombe Stream Chimpanzee Reserve, Kakombe valley, 7 Jan. 1964, *Pirozynski* 194!; Iringa, 27 Jan. 1932, *Lynes* I.h.31!

DISTR.   **T**1, 4, 7; Congo (Brazzaville), Zaire, Burundi, Malawi, Zambia, Rhodesia and Angola

HAB.   Deciduous woodland; (?550–)1000–1700 m.

NOTE.   Both *G. buchananii* Bak. and *G. huillensis* occur in the Kakombe valley. They have often been considered synonymous but I came to the conclusion long ago that two taxa were involved and am happy to follow Bamps in considering them species. Although the characters sound trivial, in practice the separation of the two is not difficult.

10.   **G. acutifolia** *N. Robson* in Bol. Soc. Brot., sér. 2, 34: 95, t. 1 (1960) & in F.Z. 1: 404 (1961). Type: Mozambique, Niassa, 2·3 km. from Muaguide, *Balsinhas* 61 (BM, holo.!, LMJ, iso.!)

Shrub or small tree 2–3 m. tall, glabrous; branchlets slender, narrowly 4-winged. Leaves opposite; blades bright green on both surfaces, ovate to lanceolate or elliptic, 6–9 cm. long, 2·5–5 cm. wide, shortly and mostly very acutely acuminate at the apex, cuneate to rounded at the base, thickly papery, venation prominent on both surfaces with 12–20 pairs of main lateral nerves and with translucent streaks and interrupted lines ± parallel to them but without visible dark longitudinal glandular lines; petiole 4–6 mm. long, channelled above, yellow-green to orange with an inconspicuous ligule. In-florescences terminal or axillary. Male flowers in few–many-flowered shortly pedunculate cymes; pedicels 1·5–4 mm. long. Sepals 4, pale green, unequal, the outer pair broadly ovate to ± round, 1–1·3 mm. long, 1–1·3 mm. wide, cucullate, the inner pair broadly obovate to round, 2–2·5 mm. long, 2 mm. wide, rounded at the apex; petals 4, yellow, oblong or oblanceolate, 3–3·5 mm. long, 1·5–2 mm. wide, rounded at the apex; stamen-bundles 2, opposite the sepals, each of 3 stamens with the filaments completely united and anthers sessile, oblong or elliptic, curved or straight, not locellate; rudimentary ovary flattened-obconic, ± 1 mm. in diameter with a 4-lobed stigma. Female flowers and fruit unknown.

TANZANIA. Uzaramo District: Pugu Forest Reserve, Aug. 1953, Semsei 1294!
DISTR. T6; Mozambique
HAB. Dry evergreen forest; 100 m.

11. **G. bifasciculata** N. *Robson* in Bol. Soc. Brot., sér. 2, 34: 94 (1960). Type: Tanzania, Morogoro District, near Kimboza on Mikese–Kisaki road, *Greenway* 2524 (K, holo.!, EA, iso.!)

Small tree or shrub 4·5–6 m. tall, glabrous; branches ± slender, at first 4-angled, at length round, smooth or rather rough; bark green, smooth. Leaves opposite; blades oblong or ovate-oblong, 7–15 cm. long, 3–7 cm. wide, obtuse or shortly and bluntly acuminate at the apex, cuneate at the base, papery; venation slightly prominent, crossed by linear gland channels; petiole 0·6–1·4 cm. long, channelled above, slender, greenish yellow or orange, without a ligule. Male flowers arranged in terminal or axillary (2–)3–10-flowered cymes, pedunculate; bracts triangular; pedicels thick, green, 2–5 mm. long, 4-angled. Sepals 4, round, 3–3·5 mm. long, rounded at the apex, the 2 outer thicker and green, the 2 inner 3·5–5·5 mm. long and resembling the petals in form and colour. Petals 4, greenish white, obovate, ± 7 mm. long. Staminal bundles 2, opposite the outer sepals, each with 3–4 oblong curved non-septate sessile or subsessile anthers; filaments ± 3 mm. long. Rudimentary ovary 2–2·5 mm. tall; stigma fleshy, punctate. Female flowers and fruits unknown.

TANZANIA. Morogoro District: near Kimboza on Mikese–Kisaki road, 4 Sept. 1930, *Greenway* 2524! & Kimboza Forest Reserve, July 1952, *Semsei* 811! & 825
DISTR. T6; not known elsewhere
HAB. Swamp forest on black soil in an area of limestone outcropping; 390 m.

Subfamily **HYPERICOIDEAE***

This group of plants was treated as Hypericaceae by E. Milne-Redhead in 1953, but it is now generally regarded as a subfamily of Guttiferae and this opportunity is taken to make certain amendments and additions.

[Subfamily description as for Hypericaceae]

6. **HYPERICUM**

L., Gen. Pl., ed. 5: 341 (1754)

[Generic description as printed]

The following sections of the genus are represented in East Africa:—

Campylosporus (Spach) R. Keller	spp. 1–4
Adenosepalum Spach	spp. 5–7
Humifusoideum R. Keller	sp. 8
Spachium (R. Keller) N. Robson	spp. 9–11

* By N. K. B. Robson.

1. Styles ± united, (4–)5; petals (1·5–)2–3·5 cm. long;
   trees or shrubs 2
   Styles free, 3–4(–5); petals 0·2–1·4 cm. long;
   shrubs or herbs 5
2. Flowers solitary; styles free at the apex; young
   shoots with internodes 2–6 mm. long 3
   Flowers in corymbose cymes; styles completely
   united; young shoots with internodes 10–15
   mm. long. 4
3. Flowers cup-shaped at anthesis; petals orange-
   yellow with bright red outside; stamen-
   filaments united for 2–5 mm.; leaf venation
   wholly parallel . . . . . 1. *H. bequaertii*
   Flowers expanded at anthesis; petals orange-
   yellow to bright yellow, very rarely tinged red
   outside; stamen-filaments united for 1 mm.
   or less; leaf venation parallel to reticulate . 2. *H. revolutum*
4. Leaves with tertiary veins forming a conspicuous
   reticulum, each areole containing a translucent glandular dot . . . . . 4. *H. roeperanum*
   Leaves without visible tertiary veins, linear translucent glands present below the upper surface 3. *H. quartinianum*
5. Black glands present on leaves, sepals, petals and
   anthers; stems 2-lined or terete; placentation
   axile 6
   Black glands completely absent; stems 4-lined;
   placentation parietal 9
6. Styles (4–)5; sepals obtuse or rounded, with intra-
   marginal black glands; fruit fleshy, indehis-
   cent . . . . 8. *H. peplidifolium*
   Styles 3(–4); sepals acute, with sessile or stalked
   marginal black glands; fruit dry, dehiscent 7
7. Stem and leaves (and whole plant) glabrous;
   much-branched shrub or weak shrublet;
   bracts without glandular auricles . . . 8
   Stem and leaves pubescent to glabrescent; erect
   perennial herb with stems usually unbranched
   below inflorescence; bracts with glandular
   auricles . . . . 7. *H. annulatum*
8. Leaves sessile, ± amplexicaul (except near base of
   shoot), oblong to elliptic-oblong, (0·5–)0·7–2
   cm. broad, cordate to rounded at the base;
   plant a slender shrub or weak shrublet with
   ± strict branches . . . . . 5. *H. conjungens*
   Leaves shortly petiolate, narrowly oblong or narrowly elliptic to obovate, 0·25–0·7(–1·0) cm.
   broad; plant a ± spreading much-branched
   shrub with ascending branches . . . 6. *H. kiboense*
9. Capsule acuminate, equalling or exceeding sepals;
   plant erect or decumbent at the base only;
   flowers in regular cymes or solitary,
   terminal . . . . . . 9. *H. lalandii*
   Capsule rounded, equalling or shorter than sepals;
   plant wholly decumbent or prostrate; flowers
   solitary, terminal or pseudo-axillary 10
10. Leaves cordate at the base, 5–7(–9)-nerved;
   sepals 5–6 mm. long; stamens 36–45 . . 10. *H. humbertii*

Leaves rounded to broadly cuneate at the base,
3–5(–7)-nerved; sepals 3–4 mm. long; sta-
mens 15–30 .   .   .   .   .   . 11. *H. scioanum*

1. **H. bequaertii** *De Wild.* in Rev. Zool. Bot. Afr. 8, Suppl. Bot.: 4 (1920) &
Pl. Bequaert. 1 : 241 (1922); F.T.E.A., Hyperic.: 5 (1953); A.V.P.: 130 (1957);
N. Robson in K.B. 12: 444 (1958); Spirlet, Guttif., Contr. Fl. Congo: 10
(1966); Bamps in F.C.B., Guttif.: 10, t. 1, fig. 1/D (1970) & in B.J.B.B. 41 :
441 (1971) & in Distr. Pl. Afr. 3, map 74 (1971)

[Description as before]

2. **H. revolutum** *Vahl*, Symb. Bot. 1 : 66 (1790); N. Robson in K.B. 14 : 251
(1960) & F.Z. 1 : 381 (1961); Spirlet, Guttif., Contr. Fl. Congo: 7, fig. 1/A–C
(1966); Moggi & Pisacchi in Webbia 22 : 236, fig. 1, map 1 (1967); Bamps in
F.C.B., Guttif.: 8, fig. 1/A, B (1970) & in B.J.B.B. 41 : 438 (1971) & in Distr.
Pl. Afr. 3, map 72 (1971); U.K.W.F.: 186 (1974); Killick & Robson in F.S.A.
22 : 15 (1976). Type: Yemen, *Forsskål* 796 (C, lecto.)

Much-branched evergreen shrub or small tree (0·3–)1–12 m. high, with
scaly bark. Young stems 4-angled, but soon becoming woody and terete.
Leaves narrowly elliptic to elliptic or narrowly oblong, ranging from
11×2·5 mm. to 60×7·5 mm., acute, narrowed to a ± broad clasping base,
furnished with 3 or 5 veins running from the base to near the apex, usually
with 1–8 cross veins and sometimes a reticulum of tertiary veins, and with
linear, often interrupted, ± translucent glands and translucent or black,
sparse or ± dense marginal glandular dots. Bracts leaf-like, with black
marginal dots. Sepals ± unequal, ovate, acute to obtuse, with or without
black marginal dots. Petals 1·5–4·3 cm. long, 0·5–2·6 cm. broad, orange-
yellow to golden-yellow, sometimes marked outside with bright red, without
black marginal dots. Stamens in 5 bundles of 30–40 each. Styles 5, united
in the lower part. Capsule 5-valved.

subsp. **revolutum**

Leaves with 3–8 cross veins and often a ± dense tertiary reticulum, interrupting
the longitudinal glandular lines. Pedicel 1–5 mm. long. Petals not marked with red.
Stamens 30–35 per fascicle, shorter than or equalling the styles.
DISTR. [Add **T4**; also in Yemen and Asir. Delete reference to Madagascar, the Comoro
Islands and Réunion]

HAB. [As before]

SYN. [*H. lanceolatum* sensu Oliv., F.T.A. 1: 156 (1868); P.O.A. C: 274 (1895); V.E.
    3(2): 497, t. 229 (1921); F.W.T.A. 1: 230 (1927); Staner in B.J.B.B. 13: 74
    (1934); T.S.K.: 36 (1936); Jex-Blake, Gard. E. Afr., ed. 2: 207, t. 3/4 (1939);
    F.P.N.A. 1: 620 (1948); F.T.E.A., Hyperic: 4 (1953); F.W.T.A., ed. 2, 1: 287,
    t. 109 (1954); N. Robson in K.B. 12: 444 (1958); Pellegr. in Bull. Soc. Bot.
    Fr. 106: 217 (1959); K.T.S.: 235, fig. 46 (1961); Mooney in Proc. Linn. Soc.
    174: 147, t. 2a (1963); Gilli in Ann. Naturhist. Mus. Wien 74: 425 (1970),
    pro parte excl. typo, *non* Lam. (1797)]
    H. leucoptychodes A. Rich., Tent. Fl. Abyss. 1: 96 (1847); Good in J.B. 65: 330,
    t. 582/3–5 (1927); Bredell in Bothalia 3: 580 (1939); Fl. Pl. S. Afr. 20, t. 787
    (1940); T.T.C.L.: 249 (1949); I.T.U., ed. 2: 157, t. 33 (1952); Pardy in Rhodes.
    Agric. Journ. 53: 514, t. 515–6 (1956). Types: Ethiopia, near Dschenausa,
    Schimper 834 & Mt. Bachit, *Schimper* 1177 (both K, isosyn.!)
    H. lanuriense De Wild., Pl. Bequaert. 5: 403 (1932); F.P.N.A. 1: 622 (1948).
    Type: Zaire, Ruwenzori, Ruanoli [Lanuri] valley, *Bequaert* 4460 (BR, holo.!)

subsp. **keniense** (*Schweinf.*) N. Robson, stat. nov.

Leaves without or with 1–2 cross veins, not interrupting the longitudinal glandular
lines, without a tertiary reticulum. Pedicel 5–8 mm. long. Petals (? always) marked
with red. Stamens ± 40 per fascicle, equalling or exceeding the styles.

DISTR. and HAB. [As before]

SYN. *H. keniense* Schweinf. in von Höhnel, Reise Zum Rudolf-See u. Stephanie-See:
15, 868 (1892); Engl., Hochgebirgsfl. Trop. Afr.: 308 (1892); P.O.A. C: 274
(1895), pro parte, quoad spec. ex Kenia; V.E. 3(2): 499 (1921), pro parte,
quoad spec. ex Kenia; Good in J.B. 65: 333, t. 582/12, 13 (1927); Milne-Redh.
in K.B. 8: 434 (1953) & in F.T.E.A., Hyperic.: 5 (1953); A.V.P. : 130 (1957);
N. Robson in K.B. 12: 444 (1958); Spirlet, Guttif., Contr. Fl. Congo: 10, fig.
1/D (1966); Bamps in F.C.B., Guttif.: 10, fig. 1/C (1970) & in B.J.B.B. 41: 440
(1971) & in Distr. Pl. Afr. 3, map 74 (1971); U.K.W.F.: 186 (1974). Type:
Kenya, Mt. Kenya, western slope at 1950 m., *von Hoehnel* (B, holo. †, BM,
fragment !)
H. ruwenzoriense De Wild. in Rev. Zool. Bot. Afr. 8, Suppl. Bot.: 5 (1920); Good
in J.B. 65: 333 (1927); Staner in B.J.B.B. 13: 71 (1934); F.P.N.A. 1: 623
(1948); T.T.C.L.: 250 (1949); I.T.U., ed. 2: 159 (1952); Spirlet, Guttif., Contr.
Fl. Congo: 8, fig. 1/E, F (1966). Type: Zaire, Ruwenzori, Butahu [Butagu]
valley, *Bequaert* 3705 (BR, holo.!)

NOTE. *H. revolutum* can be distinguished from *H. bequaertii* by its erect (not pendulous)
flowers with spreading (not erect) petals and by its more shortly united stamen-filaments. From *H. lanceolatum* Lam. (in Réunion) it differs *inter alia* in its outcurving
(not ascending) styles and in having sepals outcurving (not erect) in bud. Like *H.
lanceolatum*, however, it has differentiated into two taxa which occur at different but
overlapping altitudes and cannot always be differentiated. In both species this
apparent merging seems to have resulted from partial separation rather than from
separation and subsequent hybridisation; and so the rank of subspecies seems appro-
priate for these taxa. Subsp. *keniense* becomes less " pure " the further away one is
from the Ruwenzori Mts. and Mt. Elgon. Indeed, very few specimens from Tanzania
can be so described.

3. **H. quartinianum** *A. Rich.*, Tent. Fl. Abyss. 1 : 97 (1847); F.T.E.A.,
Hyperic.: 3 (1953); N. Robson in K.B. 12 : 444 (1958) & in F.Z. 1 : 380 (1961);
Moggi & Pisacchi in Webbia 22 : 252, fig. 5, map 3 (1967); Bamps in F.C.B.,
Guttif.: 13 (1970) & in B.J.B.B. 41 : 443 (1971) & in Distr. Pl. Afr. 3, map 76
(1971); U.K.W.F. 186 (1974)

[Description as before]

DISTR. [Add Zaire (Shaba), Zambia (Mbala), Mozambique (Niassa)]

4. **H. roeperanum** *A. Rich.*, Tent. Fl. Abyss. 1: 96 (1847); F.T.E.A.,
Hyperic.: 3 (1953); N. Robson in K.B. 12 : 444 (1958) & in F.Z. 1 : 380 (1961);
Spirlet, Guttif., Contr. Fl. Congo: 8 (1966); Moggi & Pisacchi in Webbia,
22 : 244, fig. 2, map 2 (1967), pro parte excl. subsp. *gnidiifolium*; Bamps in
F.C.B., Guttif.: 12 (1970) & in B.J.B.B. 41 : 441 (1971) & in Distr. Pl. Afr.
3, map 75 (1971); U.K.W.F. : 186 (1974); Killick & Robson in F.S.A. 22 :
16 (1976)

[Description as before]

DISTR. [Add Guinée, Nigeria, Cameroun, Zaire, Mozambique, Zambia, Rhodesia,
South Africa (Transvaal), Angola (Huila)]

SYN. *H. schimperi* A. Rich., Tent. Fl. Abyss. 1: 97 (1847). Type: Ethiopia, Ougerate,
Goumasso, *Quartin Dillon & Petit* (P, lecto.)
H. riparium A. Chev. in Mém. Soc. Bot. Fr. 8: 8 (1907); Keay & Milne-Redh.
in F.W.T.A., ed. 2, 1: 287 (1954). Type: Guinée, Ditinn, *Chevalier* 13460 (P,
holo.)
H. conrauanum Engl. in E.J. 40: 555 (1908). Type: Cameroun, Baberong, *Conrau*
28 (B, holo. †)
H. roeperanum A. Rich. var. *schimperi* (A. Rich.) Moggi & Pisacchi in Webbia
22: 249, fig. 3, map 2 (1967)

NOTE. Moggi & Pisacchi (1967) reduced *H. schimperi* to a variety, *H. roeperanum*
var. *schimperi* (A. Rich.) Moggi & Pisacchi, which they distinguished on the leaves
(shape, more discolorous, with pellucid glands more elongate) and sepals (ovate, not
ovate-oblong). The variety is said to be confined to Ethiopia, but plants resembling
it in at least some characters occur in Kenya (e.g. *Scott Elliot* 6569(K), from Gilgil).

Whether it is a variety worth maintaining is doubtful. On the other hand, the reduction by Moggi & Pisacchi of the Ethiopian endemic *H. gnidiifolium* A. Rich. to a subspecies of *H. roeperanum* does not appear to be warranted.

5. **H. conjungens** *N. Robson* in K.B. 13: 397 (1959) & in F.Z. 1: 381 (1961); Spirlet, Guttif., Contr. Fl. Congo: 6 (1966); Bamps in F.C.B., Guttif.: 5 (1970) & in B.J.B.B. 41: 436 (1971) & in Distr. Pl. Afr. 3, map 69 (1971). Type: Tanzania, Njombe District, Mdapo, *Semsei* 1643 (K, holo.!, PRE, iso.!)

Slender or bushy shrub or weak shrublet, up to 2 m. tall, with ± strict branches. Stems terete, ± woody, with the cortex soon flaking off. Leaves sessile (or sometimes very shortly petiolate towards the base of the shoot), oblong to elliptic-oblong or rarely obovate, 10–30 × (5–)7–20(–22) mm., rounded to retuse at the apex, cordate-amplexicaul to rounded at the base, with ± undulate margin, well-developed reticulate venation, numerous conspicuous translucent glandular dots and dark marginal glandular dots. Flowers numerous, in rather dense corymbose cymes. Bracts with dark marginal dots. Sepals subequal, lanceolate to oblong-lanceolate, acute, with sessile marginal dark glands and longitudinal translucent glandular streaks or lines. Petals 10–12 mm. long, primrose-yellow tinged and sometimes veined red, with marginal and subapical black dots. Stamens ± 80, 3-fasciculate or ± irregular. Styles 3, slightly longer than the ovary. Capsule 3-valved.

TANZANIA. Mbeya District: lower slopes of Poroto Mts., 18 Feb. 1934, *Michelmore* 964!; Njombe District: Matamba, 8 Jan. 1957, *Richards* 7593!
DISTR. ?**K6** (see note); **T7**; Nyika Plateau (Zambia, Malawi) and in Zaire (Marungu Mts.), probably also in Mozambique (" Mts. E. of Lake Nyasa ")
HAB. Upland grassland, grassy valleys and forest margins; 1800–2550 m.

SYN. ?*H. sp. A* sensu Milne-Redh. in F.T.E.A., Hyperic.: 12 (1953)
    H. sp. B sensu Milne-Redh. in F.T.E.A., Hyperic.: 12 (1953)
    H. conjunctum N. Robson in K.B. 12: 437, map 1 (1958); U.K.W.F.: 187 (1974), *non* Y. Kimura (1938)
    H. milne-redheadii Gilli in Ann. Naturhist. Mus. Wien 74: 425, t. 1/1 (1970). Type: Tanzania, Njombe District, Livingstone Mts., Madunda, *Gilli* 176 (W, holo.!)

VARIATION. *H. conjungens* is closely related to *H. kiboense* but is usually quite distinct in leaf shape and insertion. The population of *H. kiboense* on Mt. Meru, however, tends to have sessile leaves that are larger than usual for that species. As some specimens of *H. conjungens* have petiolate leaves towards the base of the shoot, and the range of leaf size and leaf-base shape overlap that of *H. kiboense*, the differences between these species are rather small. Their distributions, however, are distinct (but see Note).

NOTE. *Hypericum sp. A* was based on one specimen, which is labelled as having been collected in Kenya (**K6**), Masai District, dry bush north of Narok, *Rammell* in *F.D.* 3492 (K). Besides being far outside the (rather restricted) range of the rest of *H. conjungens*, this locality is ecologically quite unsuited to that species. Until the occurrence of *H. conjungens* in Kenya can be confirmed, therefore, it is advisable to regard the record with reserve.

6. **H. kiboense** *Oliv.* in Trans. Linn. Soc., ser. 2, Bot. 2: 329 (1887); F.T.E.A., Hyperic.: 6 (1953); A.V.P.: 131 (1957); N. Robson in K.B. 12: 437, 444, map 1 (1958); U.K.W.F.: 186, fig. (1974)

[Description, etc. as before]

7. **H. annulatum** *Moris*, Stirp. Sard.: 9 (1827) & Fl. Sard. 1: 323, t. 22 (1837); Milne-Redh. in K.B. 8: 435 (1953); F.T.E.A., Hyperic.: 6 (1953); N. Robson in K.B. 12: 444 (1958); Moggi & Pisacchi in Webbia 22: 272, figs. 12, 13, map 6 (1967); U.K.W.F.: 186, fig. (1974)

[Description as before, but including that of *H. afromontanum*]

DISTR. [Add central Ethiopia, also mountains of Bulgaria, Jugoslavia and Greece]
HAB. [Alter range to 1100–3600 m.]

SYN. *H. degenii* Bornm. in Magyar. Bot. Lapok 9: 90 (1910)
H. afromontanum Bullock in K.B. 1932: 492 (1932); Hook., Ic. Pl. 32, t. 3192 (1933); F.T.E.A., Hyperic.: 7, fig. 1 (1953); A.V.P.: 131 (1957); N. Robson in K.B. 12: 444 (1958); U.K.W.F.: 186 (1974)

NOTE. *H. afromontanum* appears to be no more than a higher-altitude form of *H. annulatum* in which the inflorescence is conspicuously condensed. Although such forms are known only from Mt. Elgon, they are approached in other parts of the range of *H. annulatum*, e.g. on Mt. Kenya, *C. G. Rogers* 399 (K). The European *H. montanum* L., which is a derivative of *H. annulatum*, shows a comparable range in degree of condensation of the inflorescence.

8. **H. peplidifolium** *A. Rich.*, Tent. Fl. Abyss. 1: 95 (1847); F.T.E.A., Hyperic.: 9 (1953); F.W.T.A., ed. 2, 1: 287 (1954); N. Robson in K.B. 12: 443, 445, map 2 (1958) & F.Z. 1: 383, t. 73/C (1961); Spirlet, Guttif., Contr. Fl. Congo: 4 (1966); Moggi & Pisacchi in Webbia 22: 264, fig. 9, map 5 (1967); Bamps, F.C.B., Guttif.: 3 (1970) & in B.J.B.B. 41: 433 (1971) & in Distr. Pl. Afr. 3, map 67 (1971); U.K.W.F.: 187, fig. (1974)

[Description as before]

9. **H. lalandii** *Choisy* in DC. Prodr. 1: 550 (1824); F.T.E.A., Hyperic.: 7 (1953); F.W.T.A., ed. 2, 1: 287 (1954); N. Robson in K.B. 12: 445 (1958) & in F.Z. 1: 385 (1961); Spirlet, Guttif., Contr. Fl. Congo: 6 (1966); Bamps, F.C.B., Guttif.: 5 (1970); N. Robson in Garcia de Orta, Sér. Bot. 1: 85 (1973); U.K.W.F. 187, fig. (1974); Killick & Robson in F.S.A. 22: 16 (1976)

[Description as before]

DISTR. [Delete references to Bhutan, Khasia and SW. Yunnan]

SYN. *H. pauciflorum* Kunth subsp. *involutum* (Labill.) Rödr.-Jim. in Mem. Soc. Ci. Nat. La Salle 33: 105 (1973), pro parte quoad syn. et spec. Afric. Madag.

NOTE. *H. lalandii*, although closely related to American species of the *H. pauciflorum* group, is confined to Africa, the Comoro Is. and Madagascar. The plant from the eastern Himalayas with which it has hitherto been confused is a form of *H. gramineum* G. Forster (see N. Robson in Blumea 20: 265 (1972)).

10. **H. humbertii** *Staner* in B.J.B.B. 13: 69 (1934); F.T.E.A., Hyperic.: 12 (1953); N. Robson in K.B. 12: 445 (1958); Spirlet, Guttif., Contr. Fl. Congo: 5 (1966); Bamps in F.C.B., Guttif.: 6 (1970) & in B.J.B.B. 41: 436 (1971) & in Distr. Pl. Afr. 3, map 70 (1971); N. Robson in Garcia de Orta, Sér. Bot. 1: 86 (1973)

[Description as before]

DISTR. [Add Burundi]

SYN. [*H. stolzii* sensu Milne-Redh. in K.B. 3: 455 (1948), pro parte quoad syn. *H. humbertii*]

11. **H. scioanum** *Chiov.* in Ann. Bot. Roma 9: 317 (1911); F.T.E.A., Hyperic.: 10, fig. 2 (1953); N. Robson in K.B. 12: 445 (1958) & in F.Z. 1: 386 (1961); Spirlet, Guttif., Contr. Fl. Congo: 5 (1966); Moggi & Pisacchi in Webbia 22: 278, fig. 15 (1967); Bamps in F.C.B., Guttif.: 7 (1970) & in B.J.B.B. 41: 437 (1971) & in Distr. Pl. Afr. 3, map 71 (1971); N. Robson in Garcia de Orta, Sér. Bot. 1: 86 (1973); U.K.W.F.: 187, fig. (1974)

[Description as before]

Distr. [Add Zambia, Malawi (both Nyika Plateau)]

Variation. [Delete reference to possible hybrids with *H. peplidifolium*. These species belong to distantly related sections and would not be expected to cross. No hybrids have, in fact, been recorded.]

7. VISMIA

[Add] Bamps in B.J.B.B. 36 : 428 (1966)

1. **V. orientalis** *Engl.* in V.E. 3(2) : 501 (1921) & in E. & P. Pf., ed. 2, 21 : 186 (1925); F.T.E.A., Hyperic. : 13, fig. 3 (1953); Bamps in B.J.B.B. 36 : 434, figs. 39/I, 43 (1966) & in Distr. Pl. Afr. 1, map 5 (1969); Mendes in Bol. Soc. Brot., sér. 2, 43 : 4 (1969)

[Description as before]

Distr. [Add Mozambique (Cabo Delgado)]

2. **V. pauciflora** *Milne-Redh.* in K.B. 8 : 437 (1953) & F.T.E.A., Hyperic. : 15 (1953); Bamps in B.J.B.B. 36 : 434 (1966) & in Distr. Pl. Afr. 1 : 6 (1969)

[Description as before]

8. PSOROSPERMUM

[Add] Bamps in B.J.B.B. 36 : 440 (1966)

P. febrifugum *Spach* in Ann. Sci. Nat., sér. 2, 5 : 162 (1836); F.T.E.A., Hyperic. : 16, fig. 4 (1953); F.W.T.A., ed. 2, 1 : 290 (1954); Robson in F.Z. 1 : 387 (1961); Spirlet, Guttif., Contr. Fl. Congo : 53 (1966), pro parte; Bamps in B.J.B.B. 36 : 447 (1966); Moggi & Pisacchi in Webbia 22 : 282, figs. 16, 17 (1967); Bamps in F.C.B., Guttif. : 20 (1970)

Variation. Authors in publications subsequent to F.T.E.A. Hypericaceae and F.W.T.A., ed. 2, have all regarded *P. febrifugum* as one variable species without definable varieties or subspecies, except Spirlet (B.J.B.B. 29 : 319 (1959) & Guttif., Contr. Fl. Congo (1966)), who divided it into 28 species. Bamps (B.J.B.B. 36 : 447 (1966) & F.C.B., Guttif. : 20 (1970)) reduced all Spirlet's names to synonyms of *P. febrifugum* Spach, and his circumscription would seem to include the material treated as *P. corymbiferum* and *P. lanatum* in F.T.E.A., Hyperic. : 16 (1952), though the specimens cited, *Eggeling* 1521 and 1783, are not now at EA or K.

9. HARUNGANA

[Add] Bamps in B.J.B.B. 36 : 453 (1966)

H. madagascariensis *Poir.* in Lam., Encycl. Méth. Bot. 6 : 314 (1804); F.T.E.A., Hyperic. : 19 (1953); F.W.T.A., ed. 2, 1 : 290 (1954); N. Robson in F.Z. 1 : 392 (1961); Spirlet, Guttif., Contr. Fl. Congo 14 (1966); Bamps in B.J.B.B. 36 : 453 (1966) & in F.C.B., Guttif. : 14 (1970)

[Description as before]

Note. One new combination, *Hypericum revolutum* Vahl subsp. *keniense* (Schweinf.) N. Robson, is published on p. 28.

INDEX TO GUTTIFERAE

GEOGRAPHICAL DIVISIONS OF THE FLORA

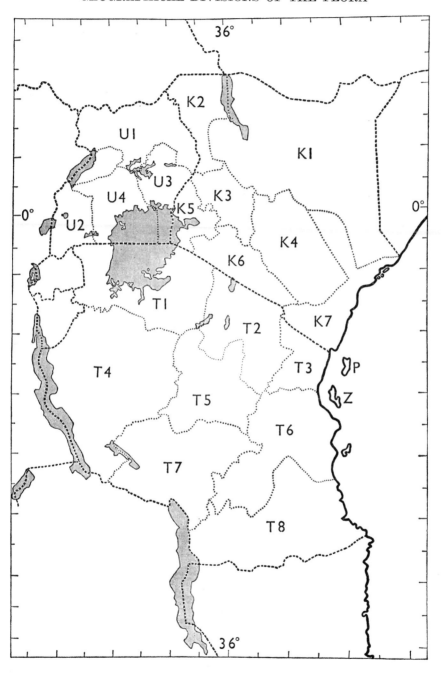